Soil Engineering

Testing, Design and Remediation

Soil Engineering

Testing, Design and Remediation

Editor

R.N. Reddy

2010

Gene-Tech Books

New Delhi - 110 002

ISBN 81-89729-96-9
ISBN 978-81-89729-96-7

Published by : **GENE-TECH BOOKS**
4762-63/23, Ansari Road, Darya Ganj,
NEW DELHI - 110 002
Phone: 41562849
e-mail: genetechbooks@yahoo.co.in

Printed at : **Chawla Offset Printers**
New Delhi – 110 052

PRINTED IN INDIA

Preface

To the engineer, the materials making up the Earth's crust are divided invariably into the categories of soil. Differing from the way an agronomist considers soil, the engineer's concern with the same lies in the fact that it more far-ranging, going beyond an agricultural necessity as the natural medium for growth of all land plants; for engineers, the term soil extends from the ground surface down to its contact with a layer of hard rock. As such, the process of soil engineering is an apt and more effective means of renewing soil resources then conventional methodology, especially as the strain on soils for the production of numerous human and animal needs continually increases.

The present text has been designed as a manual which present an incisive look into the engineering principles of soil testing, design and remediation techniques which are increasingly being used in soil conservation practice. The text attempts to acquaint readers with the essentials of the subject even as it elaborates upon the challenges, issues and concerns associated with it, relating particularly to the care of various types of soils, and how they respond to the numerous technology designed to improve their quality, texture and fertility. In addition to listing current trends and developments in soil engineering, the book offers a look at how engineering principles are changing the way soil is perceived, analysed and utilised.

R.N. Reddy

Contents

1

Concepts of Soil Engineering

Soils are considered a three-phase material composed of rock or mineral particles, water and air. The voids of a soil, the spaces in between mineral particles, contain the water and air. The engineering properties of soils are affected by four main factors: the predominant size of the mineral particles, the type of mineral particles, the grain size distribution, and the relative quantities of mineral, water and air present in the soil matrix. Fine particles (fines) are defined as particles less than 0.075 mm in diameter.

The following properties of soils are used by soil engineers in analysis of site conditions and design of earthworks, retaining structures, and foundations.

1. *Unit Weight*: Total unit weight: Cumulative weight of the solid particles, water and air in the material per unit volume. Note that the air phase is often assumed to be weightless.
2. *Dry unit weight*: Weight of the solid particles of the soil per unit volume.
3. *Saturated unit weight*: Weight of the soil when all voids are filled with water such that no air is present per unit volume. Note that this is typically assumed to occur below the water table.
4. *Porosity*: Ratio of the volume of voids (containing air and/or water) in a soil to the total volume of the soil expressed as a percentage. A porosity of 0% implies that

there is neither air nor water in the soil. Void ratio is the ratio of the volume of voids to the volume of solid particles in a soil. Void ratio is mathematically related to the porosity and is more commonly used in geotechnical formulae than porosity.

5. *Permeability*: A measure of the ability of water to flow through the soil, expressed in units of velocity.
6. *Consolidation*: As a noun, the state of the soil with regards to prior loading conditions; soils can be underconsolidated, normally consolidated or over-consolidated. As a verb, the process by which water is forced out of a soil matrix due to loading, causing the soil to deform, or decrease in volume, with time.
7. *Shear strength*: Amount of shear stress a soil can resist without failing.
8. *Atterberg Limits*: Liquid limit, plastic limit, and shrinkage limit, related to the plasticity of a soil. Used in estimating other engineering properties of a soil and in soil classification.

The soil engineer encounters a wide variety of soils in varying physical states. Because of the inherent variability of the physical properties of soil, several tests and measurements within soils engineering were developed to quantify these differences and to enable the engineer to apply the knowledge to economical design and construction.

Soil Settlement

Factors

The magnitude of a soil's settlement depends on several factors, including:

— Density.

— Void ratio.

— Grain size and shape.

— Structure.

— Past loading history of the soil deposit.

— Magnitude and method of application of the load.

— Degree of confinement of the soil mass.

Unless otherwise stated, it is assumed that the soil mass undergoing settlement is completely confined, generally by the soil that surrounds it.

Compressibility of Soil

Compressibility is the property of a soil that permits it to deform under the action of an external compressive load. Loads are primarily static loads that act, or may be assumed to act, vertically downward. Brief mention will be made of the effects of vibration in causing compression. The principal concern here is with the property of a soil that permits a reduction in thickness (volume) under a load like that applied by the weight of a highway or airfield. The compressibility of the underlying soil may lead to the settlement of such a structure.

In a general sense, all soils are compressible. That is, they undergo a greater or lesser reduction in volume under compressive static loads. This reduction in volume is attributed to a reduction in volume of the void spaces in the soil rather than to any reduction in size of the individual soil particles or water contained in the voids. If the soil is saturated before the load is applied, some water must be forced from the voids before settlement can take place. This process is called consolidation. The rate of consolidation depends on how quickly the water can escape, which is a function of the soil's permeability.

The compressibility of confined coarsegrained cohesionless soils, such as sand and gravel, is rarely a practical concern. This is because the amount of compression is likely to be small in a typical case, and any settlement will occur rapidly after the load is applied. Where these soils are located below the water table, water must be able to escape from the stratum.

In the case of coarse materials existing above the water table and under less than saturated conditions, the application of a static load results in the rearrangement of soil particles. This produces deformation without regard to moisture escape. So, generally speaking, settlement occurs during the period of load application. Deformations that are thus produced in sands and gravels are essentially permanent in character. There is little tendency for the soil to return to its original dimensions or rebound when the load is removed.

A sand mass in a compact condition may eventually attain some degree of elasticity with repeated applications of load. The compressibility of a loose sand deposit is much greater than that of the same sand in a relatively dense condition. Generally, structures should not be located on loose sand deposits. Avoid loose sand deposits if possible, or compact to a greater density before the load is applied. Some cohesionless soils, including certain very fine sands and silts, have loose structures with medium settlement characteristics. Both gradation and grain shape influence the compressibility of a cohesionless soil.

Gradation is of indirect importance in that a well-graded soil generally has a greater natural density than one of uniform gradation. Soils that contain platy particles are more compressible than those composed entirely of bulky grains. A fine sand or silt that contains mica flakes maybe quite compressible. Although soils under static loads are emphasized here, the effects of vibration should also be mentioned.

Vibration during construction may greatly increase the density of cohesionless soils. A loose sand deposit subjected to vibration after construction may also change to a dense condition. The latter change in density may have disastrous effects on the structures involved. Cohesionless soils are usually compacted or "densified" as a planned part of construction operations. Cohesive soils are usually insensitive to the effects of vibrations.

Soil Consolidation

Consolidation is the time-dependent change in volume of a soil mass under compressive load that occurs when water slowly escapes from the pores or voids of the soil. The soil skeleton is unable to support the load and changes structure, reducing its volume and producing vertical settlement.

Cohesive Soils

The consolidation of cohesive, fine-grained soils (particularly clays) is quite different from the compression of cohesionless soils. Under comparable static loads, the consolidation of a clay may be much greater than coarsegrained soils and settlement may take a very long time to occur. Structures often settle due to consolidation of a saturated clay stratum. The consolidation of thick, compressible clay layers is serious and may cause structural damage.

In uniform settlement, the various parts of a structure settle approximately equal amounts. Such uniform settlement may not be critical. Nonuniform, or differential, settlement of parts of a structure due to consolidation causes serious structural damage. A highway or airfield pavement may be badly damaged by the nonuniform settlement of an embankment founded on a compressible soil.

The consolidation characteristics of a compressible soil should be determined for rational design of many large structures founded on or above soils of this type. Consolidation characteristics generally are determined by laboratory consolidation tests performed on undisturbed samples.

The natural structure, void ratio, and moisture content are preserved as carefully as possible for undisturbed samples. However, military soils analysts are not equipped or trained to perform consolidation tests. Information on consolidation tests and settlement calculations for the design of structures to be built on compressible soils may be found in Naval Facilities Engineering Command Design Manual.

Shearing Resistance of Soil

From an engineering viewpoint, one of the most important properties a soil possesses is shearing resistance or shear strength. A soil's shearing resistance under given conditions is related to its ability to withstand loads. The shearing resistance is especially important in its relation to the supporting strength, or bearing capacity, of a soil used as abase or subgrade beneath a road, runway, or other structure. The shearing resistance is also important in determining the stability of the slopes used in a highway or airfield cut or embankment and in estimating the pressures exerted against an earthretaining structure, such as a retaining wall.

Test Procedures

Three test procedures are commonly used in soil mechanics laboratories to determine the shear strength of a soil. These are:

— Direct shear test.

— Triaxial compression test.

— Unconfined compression test.

Military soils analysts are not equipped or trained to perform the direct shear or triaxial compression tests. For most military applications, the CBR value of a soil is used as a measure of shear strength. A variation of the unconfined compression test can be performed by military soils analysts, but the results are ordinarily used only in evaluation of soil stabilisation. Shear strength for a soil is expressed as a combination of an apparent internal angle of friction (normally associated with cohesive soils).

California Bearing Ratio (CBR)

The CBR is a measure of the shearing resistance of a soil under carefully controlled conditions of density and moisture. The CBR is determined by a penetration shear test and is used with empirical curves for designing flexible pavements. Recommended design procedures for flexible pavements are presented in TM 5-330. The CBR test procedure for use in design consists of the following steps:

— Prepare soil test specimens.

— Perform penetration test on the prepared soil samples.

— Perform swell test on soil test specimens.

Although a standardised procedure has been established for the penetration portion of the test, one standard procedure for the preparation of test specimens cannot be established because soil conditions and construction methods vary widely. The soil test specimen is compacted so it duplicates as nearly as possible the soil conditions in the field.

In a desert environment, soil maybe compacted and tested almost completely dry. In a wet area, soil should probably be tested at 100 percent saturation. Although penetration tests are most frequently performed on laboratory-compacted test specimens, they may also be performed on undisturbed soil samples or on in-place soil in the field. Values of the field CBR may range from as low as 3 for highly plastic, inorganic clays (CH) and some organic clays and silts (OH) to as high as 80 for well-graded gravel and gravel-sand mixtures.

Airfield Index (AI)

Engineering personnel use the airfield cone penetrometer to determine an index of soil strengths (called Airfield Index) for various military applications.

Airfield Cone Penetrometer

The airfield cone penetrometer is compact, sturdy, and simple enough to be used by military personnel inexperienced in soil strength determination. If used correctly, it can serve as an aid in maintaining field control during construction operations; however, this use is not recommended, because more accurate methods are available for use during construction.

Description: The airfield cone penetrometer is a probe-type instrument consisting of a right circular cone with a base diameter of ½ inch mounted on a graduated staff. On the opposite end of the staff are a spring, a load indicator, and a handle. The overall length of the assembled penetrometer is 36 1/8 inches. For ease in carrying, the penetrometer can be disassembled into three main pieces. They are:

— Two extension staffs, each 12 5/8 inches long.

— One piece 14 ¾ inches long containing the cone, handle, spring, and load indicator.

The airfield cone penetrometer has a range of zero to 15. The airfield cone penetrometer must not be confused with the trafficability penetrometer, a standard military item included in the Soil Test Set.

The cone penetrometer used for trafficability has a dial-type load indicator (zero to 300 range) and is equipped with a cone ½ inch in diameter and a cross-sectional area of 0.2 square inch and another cone 0.8 inch in diameter and a crosssectional area of 0.5 square inch. If the trafficability penetrometer is used to measure the AI, the readings obtained with the 0.2-square-inch cone must be divided by 20; the reading obtained with the 0.5-squareinch cone must be divided by 50.

Operation: Before the penetrometer is used, inspect the instrument to see that all joints are tight and that the load indicator reads zero. To operate the penetrometer, place your palms down symmetrically on the handle. Steady your arms against your thighs and apply force to the handle until a slow, steady, downward movement of the instrument occurs.

Read the load indicator at the moment the base of the cone enters the ground and at desired depths at the moment the corresponding depth mark on the shaft reaches the soil surface. The reading is made by shifting the line of vision from the soil surface to the indicator just a moment before the desired depth is reached. Maximum efficiency is obtained with a two-person team in which one person operates and reads the instrument while the other acts as a recorder. One person can operate the instrument and record the measurements by stopping the penetration at any intermediate depth, recording previous readings, and then resuming penetration. Observe the following rules to obtain accurate data:

— Make sure the instrument reads zero when suspended by the handle and 15 when a 150-pound load is applied.

— Keep the instrument in a vertical position while it is in use.
— Control the rate of penetration at about ½ to 1 inch per second. (Slightly faster or slower rates will not materially affect the readings, however.)
— If you suspect the cone is encountering a stone or other foreign body at the depth where a reading is desired, make another penetration nearby.
— Take readings at the proper depths. (Carelessness in determining depth is one significant source of error in the use of the penetrometer.)

Maintenance: The airfield cone penetrometer is simply constructed of durable metals and needs little care other than cleaning and oiling. The calibration should be checked occasionally. If an error in excess of about 5 percent is noted, recalibrate the penetrometer.

Evaluation of Soil-strength

The number of measurements to be made, the location of the measurements, and other such details vary with each area to be examined and with the time available. For this reason, hard and fast rules for evaluating an airfield are not practical, but the following instructions are useful:

Fine-grained soils

A reading near zero can occur in a very wet soil; it cannot support traffic. A reading approaching 15 occurs in dry, compact clays or silts and tightly packed sands or gravels. Most aircraft that might be required to use an unpaved area could easily be supported for a substantial number of landing and takeoffs on a soil having an AI of 15. Soil conditions are extremely variable. As many penetrometer measurements should be taken as time and circumstances permit.

The strength range and uniformity of an area controls the number of measurements necessary. Areas obviously too soft for emergency landing strips will be indicated after a few measurements, as will areas with strengths that are more than adequate. In all areas, the spots that appear to be softest should be tested first, since the softest condition of an area controls suitability. Soft spots are not always readily apparent. If the first test results indicate barely adequate strength, the entire area should be examined.

Penetrations in areas that appear to be firm and uniform may be few and widely spaced, perhaps every 50 feet along the proposed centerline. In areas of doubtful strength, the penetrations should be more closely spaced, and areas on both sides of the centerline should be investigated. No fewer than three penetrations should be made at each location and usually five are desirable. If time permits, or if inconsistencies are apparent, as many as 10 penetrations should be made at each test location.

Soil strength usually increases with depth; but in some cases, a soil has a thin, hard crust over a deep, soft layer or has thin layers of hard and soft material. For this reason, each penetration should be made to a 24-inch depth unless prevented by a very firm condition at a lesser depth. When penetration cannot be made to the full 24-inch depth, a hole should be dug or augured through the firm materials, and penetrometer readings should be taken in the bottom of the hole to ensure that no soft layer underlies the firm layer. If possible, readings should be taken every 2 inches from the surface to a depth of 24 inches.

Generally, the surface reading should be disregarded when figures are averaged to obtain a representative AI. In the normal soil condition, where strength increases with depth, the readings at the 2- to 8-inch depths (4 to 10 inches for dry sands and for larger aircraft) should be used to designate the soil strength for airfield evaluation. If readings in this critical layer at any one test location do not differ more than 3 or 4 units, the arithmetic average of these readings can be taken as the AI for the areas represented by the readings.

When the range between the highest and lowest readings is more than about 4, the engineer must use judgment in arriving at a rating figure. For conservatism, the engineer should lean toward the low readings. In an area in which hard crust less than about 4 inches thick overlies a much softer soil, the readings in the crust should not be used in evaluating the airfield. For example, if a 3-inch-thick crust shows average readings of 10 at the 2-inch depth and average readings of 5 below 3 inches, the area should be evaluated at 5.

If the crust is more than about 4 inches thick, it will probably play an important part in aircraft support. If the crust in the above instance is 5 inches thick, the rating of the field would then be about halfway between the 10 of the crust and the 5 of the underlying soil or, conservatively, 7. Innumerable combinations of crust thickness and strength and underlying soil strength can occur. Sound reasoning and engineering judgment should be used in evaluating such areas.

In an area in which a very soft, thin layer is underlain by a firmer layer, the evaluation also is a matter of judgment. If, for example, there are 1 to 2 inches of soil with an index averaging about 5 overlying a soil with an index of 10, rate the field at 10; but if this soft layer is more than about 4 inches thick, rate the field at 5. Areas of fine-grained soils with very low readings in the top 1 inch or more are likely to be slippery or sticky, especially if the soil is a clay.

Coarse-grained soils

When relatively dry, many sands show increasing AIs with depth, but the 2-inch depth index will often be low, perhaps about 3 or 4, Such sands usually are capable of supporting aircraft that require a much higher AI than 3 to 4, because the strength of the sand actually increases under the confining action of the aircraft tires. Generally, any dry sand or gravel is adequate for aircraft in the C-130 class, regardless of the penetrometer readings. All sands and gravel in a "quick" condition must he avoided.

Evaluation of wet sands should be based on the penetrometer readings obtained as described earlier. Once the strength of the soil, in terms of AI, has been established by use of the airfield cone penetrometer, the load-carrying capability of this soil can be determined for each kind of forward, support, or rear-area airfield through use of the subgrade strength requirements curves. These curves are based on correlations of aircraft performance and AIs.

Unfortunately, these are not exact correlations uniquely relating aircraft performance to AI. As soils vary in type and condition from site to site, so varies the relation of AI to aircraft performance. For this reason, the curves may not accurately reflect performance in all cases. These relations were selected so that in nearly all cases aircraft performance will be equal to or better than that indicated.

Correlation between CBR and AI

Expedient soil strength measurements in this manual are treated in terms of AI. Measurement procedures using the airfield penetrometer are explained; however, in the references listed at the end of this manual, which cover less expedient construction methods, soil strength is treated in terms of CBR.

This correlation has been established to yield values of CBR that generally are conservative. The tendency toward conservatism is necessary because there is no unique relationship between these measurements over a wide range of soil types.

Bearing Capacity of a Soil

The bearing capacity of a soil is its ability to support loads that may be applied to it by an engineering structure, such as:

- — A building, a bridge, a highway pavement, or an airport runway and the moving loads that may be carried thereon.
- — An embankment.
- — Other types of load.

A soil with insufficient bearing capacity to support the loads applied to it may simply fail by shear, allowing the structure to move or sink into the ground. Such a soil may fail because it undergoes excessive deformation, with consequent damage to the structure. Sometimes the ability of a soil to support loads is simply called its stability. Bearing capacity is directly related to the allowable load that may be safely placed on a soil. This allowable load is sometimes called the allowable soil pressure.

Failure may take place on a number of surfaces within the soil, usually accompanied by bulging of the ground around the foundation. The ultimate bearing capacity not only is a function of the nature and condition of the soil involved but also depends on the method of application of the load.

Principle Function of a Foundations

The principle function of a foundation is to transmit the weight of a structure and the loads that it carries to the underlying soil or rock. A foundation must be designed to be safe against a shear failure in the underlying soil. This means that the load placed on the soil must not exceed its ultimate bearing capacity.

Shallow Foundations

A shallow foundation is one that is located at, or slightly below, the surface of the ground. A typical foundation of this type is seen in the shallow footings, either of plain or reinforced concrete, which may support a building. Footings are generally square or rectangular. Long continuous or strip footings are also used, particularly beneath basement or retaining walls. Another type of shallow foundation is the raft or mat; it may cover a large area, perhaps the entire area occupied by a structure.

Deep Foundations

When the surface soils at the site of a proposed structure are too weak and compressible to provide adequate support, deep foundations are frequently used to transfer the load to underlying suitable soils. Two common types of deep foundations are:

— Pile
— Pier

Pile foundations

Piles and pile foundations are very commonly used in both military and civil construction. By common usage, a pile is a load-bearing member made of timber, concrete, or steel, which is generally forced into the ground. Piles are used in a variety of forms and for a variety of purposes. A pile foundation is one or more piles used to support a pier, or column, or a row of piles under a wall. Piles of this type are normally used to support vertical loads, although they may also be used to support inclined or lateral forces.

Piles driven vertically and used for the direct support of vertical loads are commonly called bearing piles. They may be used to transfer the load through a soft soil to an underlying firm stratum. These are called end-bearing piles. Bearing piles may also be used to distribute the load through relatively soft soils that are not capable of supporting concentrated surface loads. These are called friction piles.

A bearing pile may sometimes receive its support from a combination of end bearing and friction. Bearing piles also may be used where a shallow foundation would likely be undetermined by scour, as in the case of bridge piers. A typical illustration of an end-bearing pile is when a pile driven through a very soft soil, such as a loose silt or the mud of a river bottom, comes to rest on firm stratum beneath. The firm stratum may, for example, be rock, sand, or gravel. In such cases, the pile derives practically all its support from the underlying firm stratum.

A friction pile develops its load-carrying capacity entirely, or principally, from skin friction along the sides of the pile. The load is transferred to the adjoining soil by friction between the pile and the surrounding soil. The load is thus transferred downward and laterally to the soil. The soil surrounding the pile or group of piles, as well as that beneath the points of the piles, is stressed by the load. Some piles carry a load by a combination of friction and end bearing.

A pile of this sort may pass through a fairly soft soil that provides some frictional resistance; then it may pass into a firm layer that develops loadcarrying capacity through a combination of friction over a relatively short length of embedment and end bearing. Piles are used for many purposes other than support for vertical loads. Piles that are driven at an angle with the vertical are commonly called batter piles. They may be used to support inclined loads or to provide lateral loads.

Piles are sometimes used to support lateral loads directly, as in the pile fenders that may be provided along waterfront structures to take the wear and shock of docking ships. Sometimes piles are used to resist upward, tensile forces. These are frequently called anchor piles. Anchor piles may be used, for example, as anchors for bulkheads, retaining walls, or guy wires. Vertical piles are sometimes driven for the purpose of compacting loose cohesionless deposits. Closely spaced piles, or sheet piles, may be driven to form a wall or a bulkhead that restrains a soil mass.

Piers

Piers are much less common than piles and are normally used only for the support of very heavy loads.

Earthretaining Structures

Earthretaining structures must be used to restrain a mass of earth that will not stand unsupported. Such structures are commonly required when a cut is made or when an embankment is formed with slopes too steep to stand alone. Earthretaining structures are subjected to lateral thrust from the earth masses that they support.

The pressure of the earth on such a structure is commonly called lateral earth pressure. The lateral earth pressure that may be exerted by a given soil on a given structure is a function of many variables. It must be estimated with a reasonable degree of accuracy before an earthretaining structure may be properly designed. In many cases, the lateral earth pressure may be assumed to be acting in a horizontal direction or nearly so.

Types of Earthretaining Structures

Retaining Walls

A retaining wall is a wall constructed to support a vertical or nearly vertical earth bank that, in turn, may support vertical loads. Generally, retaining walls are classified into the following five types:

— Gravity.

— Cantilever.

— Counterfort.

— Buttressed.

— Crib.

When a retaining wall is used to support the end of a bridge span, as well as retain the earth backfill, it is called an abutment. There are several types of gravity retaining walls, such as:

— Timber.

— Plain concrete.

— Sheet piling.

— Rubble.

— Stone or brick masonry.

— Crib.

— Gabions.

Retaining walls are used in many applications. For example, a structure of this sort may be used in a highway or railroad cut to permit the use of a steep slope and avoid excessive amounts of excavation. Retaining walls are similarly used on the embankment side of sidehill sections to avoid excessive volumes of fill. Bridge abutments and the headwalls of culverts frequently function as retaining walls.

In the construction of buildings and various industrial structures, retaining walls are often used to provide support for the side of deep, permanent excavations. Permanent retaining walls are generally constructed from plain or reinforced concrete; stone masonry walls are also used occasionally. In military construction, timber crib retaining walls are important.

Design of the backfills

The design of the backfill for a retaining wall is as important as the design of the wall itself. The backfill must be materials that are:

— Reasonably clean.

— Granular.

— Essentially cohesionless.

— Easily drained.

— Not susceptible to frost action.

The best materials for backfills behind retaining walls are clean sands, gravels, and crushed rock. In the USCS, the (GW) and (SW) soils are preferred, if they are available. The (GP) and (SP) soils are also satisfactory. These granular materials require compaction to make them stable against the effects of vibration. Compaction also generally increases the angle of internal friction, which is desirable in that it decreases the lateral pressure exerted on the wall.

Materials of the (GM), (GC), (SM), and (SC) groups maybe used for backfills behind retaining walls, but they must be protected against frost action and may require elaborate drainage provisions. Fine- grained soils are not desirable as backfills because they are difficult to drain. If clay soil must be used, the wall should be designed to resist earth pressures at rest. Ideal backfill materials are purely granular soils containing < 5 percent of fines.

Backfills behind retaining walls are commonly put in place after the structure has been built. The method of compaction depends on the:

— Soil.

— Equipment available.

— Working space.

Since most backfills are essentially cohesionless, they are best compacted by vibration. Common practice calls for the backfill to be placed in layers of loose material that, when compacted, results in a compacted layer thickness of from 6 to 8 inches. Each layer is compacted to a satisfactory density. In areas inaccessible to rollers or similar compacting equipment, compaction may be done by the use of mechanical air tampers or hand tools.

Drainage of the backfill

Drainage of the backfill is essential to keep the wall from being subjected to water pressure and to prevent frost action.

When the backfill is composed of clean, easily drained materials, it is customary to provide for drainage by making weep holes through the wall. Weep holes are commonly made by embedding pipes 4 to 6 inches in diameter into the wall. These holes are spaced from 5 to 10 feet center to center both horizontally and vertically. A filter of granular material should be provided around the entrance to each weep hole to prevent the soil from washing out or the drain from becoming clogged. If possible, this material should conform to the requirements previously given for filter materials.

Weep holes have the disadvantage of discharging the water that seeps through the backfill at the toe of the wall where the soil pressures are greatest. The water may weaken the soil at this point and cause the wall to fail. A more effective solution, which is also more expensive, is to provide a longitudinal back drain along the base of the wall. A regular pipe drain should be used, surrounded with a suitable filter material.

The drainage may be discharged away from the ends of the wall. If a granular soil, which contains considerable fine material and is poorly drained (such as an (SC) soil) is used in the backfill, then more elaborate provisions may be installed to ensure drainage. One such approach is to use a drainage blanket. If necessary, a blanket of impervious soil or bituminous material may be used on top of the backfill to prevent water from entering the fill from the top. Such treatments are relatively expensive.

Frost action

Conditions for detrimental frost action on retaining walls include the following:

— A frost-susceptible soil.

— Availability of water.

— Freezing temperatures.

If these conditions are present in the backfill, steps must be taken to prevent the formation of ice lenses and the resultant severe lateral pressures that may be exerted against the wall. The usual way to prevent frost action is to substitute a thick layer of clean, granular, nonfrost-susceptible soil for the backfill material immediately adjacent to the wall. The width of the layer should be as great as the maximum depth of frost penetration in the area. As with other structures, the bottom of a retaining wall should be located beneath the line of frost penetration.

Timber crib

A very useful type of retaining wall for military purposes in a theater of operations is timber cribbing. The crib, or cells, are filled with earth, preferably clean, coarse, granular material. A wall of this sort gains its stability through the weight of the material used to fill the cells, along with the weight of the crib units themselves. The longitudinal member in a timber crib is called a stretcher, while a transverse member is a header.

A principal advantage of a timber crib retaining wall is that it may be constructed with unskilled labor and a minimum of equipment. Suitable timber is available in many military situations. Little foundation excavation is usually required and may be limited to shallow trenching for the lower part of the crib walls. The crib maybe built in short sections, one or two cribs at a time. Where the amount of excavation is sufficient and suitable, the excavated soil may be used for filling the cells.

A crib of this sort maybe used on foundation soils that are weak and might not be able to support a heavy wall, since the crib is fairly flexible and able to undergo some settlement and shifting without distress. However, this should not be misunderstood, as the foundation soil must not be so soft as to permit excessive differential settlement that would destroy the alignment of the crib.

Experience indicates that a satisfactory design will generally be achieved if the base width is a minimum of 4 feet at the top and bottom or 50 percent of the height of the wall, provided that the wail does not carry a surcharge and is on a reasonably firm foundation. If the wall carries a heavy surcharge, the base width should be increased to a minimum of 65 percent of the height. In any case, the width of the crib at the top and bottom should not be less than 4 feet.

Timber crib walls may be built with any desired batter (receding upward slope) or even vertical. The batter most often used and recommended is one horizontal to four vertical. If less batter is used, the base width must be increased to ensure that the resultant pressure falls within the middle third of the base. The desired batter is normally achieved by placing the base on a slope equal to the batter. The toe may be placed on sills; this is frequently done with high walls. Sometimes double-cell construction is used to obtain the necessary base width of high walls.

The wall is then decreased in width, or "stepped-back," in the upper portions of the wall, above one third height. Additional rows of bottom stretchers may be used to decrease the pressure on the soil or to avoid detrimental settlement. The front and rear wall of the crib should be connected at each panel point.

The crib must be kept an essentially flexible structure and must be free to move somewhat in any direction, so as to adjust itself to thrusts and settlements. The material used in filling the cells should be placed in thin layers and should be well compacted. Backfill behind the wall should also be compacted and kept close to, but not above, the level of the material in the cribs. Drainage behind timber crib walls may or may not be required, depending on local conditions and wall construction.

Other types of timber walls

Other types of timber retaining walls are used for low heights, particularly in connection with culverts and bridges. A wall of this sort maybe built by driving timber posts into the ground and attaching planks or logs.

Gabions

Gabions are large, steel-wiremesh baskets usually rectangular in shape and variable in size. They are designed to solve the problem of erosion at a low cost. Gabions were used in sixteenth century fortifications, and experience in construction with factory-produced prefabricated gabions dates back to 1894 in Italy.

Gabions have been wideiy used in Europe and are now becoming accepted in the United States as a valuable and practical construction tool. They can be used in place of sheet piling, masonry construction, or cribbing. Gabions maybe used as:

— Protective and antierosion structures on rivers (as revetments, groynes, or spurs).

— Channel linings.

— Seashore protection.

— Retaining walls for roads or railroads.

— Antierosion structures (such as weirs, drop structures, and check dams).

— Low-water bridges or fords.

— Culvert headwall and outlet structures.

— Bridge abutments and wing walls.

The best use of gabions as retaining walls is where flexibility and permeability are important considerations, especially where unstable ground and drainage conditions impose problems difficult to solve with rigid and impervious material. Use of gabions does require ready access to large-size stones, such as those found in mountainous areas.

Areas that are prone to landslides have used gabions successfully. Gabion walls have been erected in mountainous country to trap falling rocks and debris and in some areas to act as longitudinal drainage collectors. The best filling material for a gabion is one that allows flexibility in the structure but also fills the gabion compartments with the minimum of voids and with the maximum weight. Ideally, the stone should be small, just slightly larger than the size of the mesh. The stone must be clean, hard, and durable to withstand abrasion and resistance to weathering and frost action.

The gabions are filled in three lifts, one foot at a time. Rounded stone, if available, reduces the possibility of damage to the wire during mechanical filling as compared with sharp quarry stone. If stone is not available, gabions can be filled with a good quality soil. To hold soil, hardware cloth inserts must be placed inside the gabions. For use in gabions, backfill material should meet the following Federal Highway Administration criteria:

— For a 6-inch sieve, 100 percent of the material should pass through.
— For a 3-inch sieve, 75 to 100 percent of the material should pass through.
— For a Number 200 sieve, zero to 25 percent of the material should pass through.
— The PI should be 6 or less.

Bracing Systems

Bracing systems maybe required to protect the sides of temporary excavations during construction operations. Such temporary excavations may be required for several purposes but are most often needed in connection with the construction of foundations for structures and the placing of utility lines, such as sewer and water pipes.

Shallow excavations

The term "shallow excavation" refers to excavations made to depths of 12 to 20 feet below the surface, depending principally on the soil involved. The lower limit applies to fairly soft clay soils, while the upper limit generally applies to sands and sandy soils.

Shallow excavations may be made as open cuts with unsupported slopes, particularly when the excavation is being done above the water table. If the excavation is purely temporary in nature, most sandy soils above the water table will stand at somewhat steeper slopes (as much as ½ to 1 for brief periods), although some small slides may take place. Clays may be excavated to shallow depths with vertical slopes and will remain stable briefly. Generally, bracing cuts in clay that extend to depths of 5 feet or more below the surface are safer unless flat slopes are used.

Even for relatively shallow excavations, using unsupported cuts may be unsatisfactory for several reasons. Cohesive soils may stand on steep slopes temporarily, but bracing is frequently needed to protect against a sudden cave-in. Required side slopes, particularly in loose, granular soils, may be so flat as to require an excessive amount of excavation. If the excavation is being done close to other structures, space maybe limited, or the consequences of the failure of a side slope may be very serious.

Considerable subsidence of the adjacent ground may take place, even though the slope does not actually fail. Finally, if the work is being done below the water table, the excavation may have to be surrounded with a temporary structure that permits the excavation to be unwatered.

Narrow shallow excavations

Several different schemes may be used to brace the sides of a narrow shallow excavation. In the first scheme, timber planks are driven around the boundary of the excavation to form what is called vertical sheeting. The bottom of the sheeting is kept at or near the bottom of the pit or trench as excavation proceeds. The sheeting is held in place by means of horizontal beams called wales. These wales are usually supported against each other by means of horizontal members called struts, which extend from one side of the excavation to the other.

The struts may be cut slightly long, driven into place, and held by nails or cleats. They may also be held in position by wedges or shims. Hydraulic or screw-type jacks can be used as struts. The second scheme uses horizontal timber planks to form what is called horizontal lagging. The lagging, in turn, is supported by vertical solid beams and struts. If the excavation is quite wide, struts may have to be braced horizontally or vertically or both.

Bracing systems for shallow excavations are commonly designed on the basis of experience. Systems of this sort represent cases of incomplete deformation, since the bracing system prevents deformation at some points while permitting some deformation at others. Members used in bracing systems should be strong and stiff.

In ordinary work, struts vary from 4-inch to 6-inch timbers for narrow cuts up to 8-inch by 8-inch timbers for excavations 10 or 12 feet wide. Heavier timbers are used if additional safety is desired. Struts are commonly spaced about 8 feet horizontally and from 5 to 6 feet vertically. Lagging or sheeting is customarily made from planks from 6 to 12 inches wide, with the minimum thickness usually being about 2 inches.

Wide shallow excavations

If the excavation is too wide to be cross braced by the use of struts, vertical sheeting may be used. The wales are supported by inclined braces, which are sometimes called rakes. The rakes, in turn, react against kicker blocks that are embedded in the soil. As the excavation is deepened, additional wales and braces may be added as necessary to hold the sheeting firmly in position. The success of this system depends on the soil in the bottom of the excavation being firm enough to provide adequate support for the blocks.

Soil Behaviour and Civil Engineering

Civil engineering includes the conception, analysis, design, construction, operation, and maintenance of a diversity of structures, facilities, and systems. All are built on, in, or with soil or rock. The properties and behaviour of these materials have major influences on the success, economy, and safety of the work. Geoengineers play a vital role in these projects and are also concerned with virtually all aspects of environmental control, including water resources, water pollution control, waste disposal and containment, and the mitigation of such natural disasters as floods, earthquakes, landslides, and volcanoes.

Soils and their interactions with the environment are major considerations. Furthermore, detailed understanding of the behaviour of earth materials is essential for mining, for energy resources development and recovery, and for scientific studies in virtually all the geosciences. To deal properly with the earth materials associated with any problem and project requires knowledge, understanding, and appreciation of the importance of geology, materials science, materials testing, and mechanics.

Geotechnical engineering is concerned with all of these. Environmental concerns-especially those related to groundwater, the safe disposal and containment of wastes, and the cleanup of contaminated sites-has spawned yet another area of specialisation; namely, environmental geotechnics, wherein chemistry and biological science are important. Geochemical and microbiological phenomena impact the composition, properties, and stability of soils and rocks to degrees only recently beginning to be appreciated.

Students in civil engineering are often quite surprised, and sometimes quite confused, by their first course in engineering with soils. After studying statics, mechanics, and structural analysis and design, wherein problems are usually quite clear-cut and well defined, they are suddenly confronted with situations where this is no longer the case. A first course in soil mechanics may not, at least for the first half to two-thirds of the course, be mechanics at all. The reason for this is simple:

Analyses and designs are useless if the boundary conditions and material properties are improperly defined. Acquisition of the data needed for analysis and design on, in, and with soils and rocks can be far more difficult and uncertain than when dealing with other engineering materials and aboveground construction. There are at least three important reasons for this.

— *No Clearly Defined Boundaries*: The free body of the cantilever beam is readily analysed for reactions, shears, moments, and deflections using standard methods of structural analysis.

— *Variable and Unknown Material Properties*: The properties of most construction materials are ordinarily known within rather narrow limits and usually can be specified to meet certain needs. Although this may be the case in construction using earth and rock fills, at least part of every geotechnical problem involves interactions with in situ soil and rock. No matter how extensive any boring and sampling program, only a very small percentage of the subsurface material is available for observation and testing. In most cases, more than one stratum is present, and conditions are nonhomogeneous and anisotropic.

— *Stress and Time-Dependent Material Properties*: Soils, and also some rocks, have mechanical properties that depend on both the stress history and the present stress state. This is because the volume change, stress-strain, and strength properties depend on stress transmission between particles and particle groups. These stresses are, for the most part, generated by body forces and boundary stresses and not by internal forces of cohesion, as is the case for many other materials. In addition, the properties of most soils change with time after placement, exposure, and loading. Because of these stress and time dependencies, any given geotechnical problem may involve not just one or two but an almost infinite number of different materials.

Add to the above three factors the facts that soil and rock properties may be susceptible to influences from changes in temperature, pressure, water availability, and chemical and biological environment, and one might conclude that successful application of mechanics to earth materials is an almost hopeless proposition. It has been amply demonstrated, of course, that such is not the case; in fact, it is for these very reasons that geotechnical engineering offers such a great challenge for imaginative and creative work.

Modern theories of soil mechanics, the capabilities of modern computers and numerical analysis methods, and our improved knowledge of soil physics and chemistry make possible the solution of a great diversity of static and dynamic problems of stress deformation and stability, the transient and steady-state flow of fluids through the ground, and the long-term performance of earth systems. Nonetheless, our ability to analyse and compute often exceeds considerably our ability to understand, measure, and characterise a problem or process. Thus, understanding and the ability to conceptualise soil and rock behaviour become all the more important.

Consideration of Soil Formation

Water may make up more than half the volume of a soil mass, it is attracted to soil particles, and the interactions between water and the soil surfaces influence the behaviour. In addition, owing to the colloidal nature of clay particles, the types and concentrations of chemicals in a soil can influence significantly its behaviour in a variety of ways.

References

Bowles, J., *Foundation Analysis and Design*, McGraw-Hill Publishing Company, 1988.

Bray, R.H., "Requirements for successful soil tests," *Soil Sci.*, 1948, 66(2):83-89.

Cedergren, Harry R., *Seepage, Drainage, and Flow Nets*, Wiley, 1977.

Luebs, R.E., G. Stanford, and A.D. Scott., "Relation of available potassium to soil moisture," *Soil Sci. Soc. Amer. Proc.*, 1956, 20:45-50.

Schofield, R. K., and A. W. Taylor., "The measurement of soil pH," *Soil Sci. Soc. Amer.*, Proc.19:164-167, 1955.

Spangler, M.G. and R.L. Handy (1982). *Soil Engineering*. New York: Harper & Row. 819 pp

Terzaghi, K., Peck, R.B. and Mesri, G., *Soil Mechanics in Engineering Practice* 3rd Ed., John Wiley & Sons, Inc., 1996.

Thien, S.J., D.A. Whitney, and D.L. Karlen., "Effect of microwave radiation drying on soil chemical and mineralogical analysis," *Comm. Soil Sci. Plant Anal.*, 1978, 9:231-241.

2

Soil Mechanics

Soil mechanics is a discipline that applies principles of engineering mechanics, e.g. kinematics, dynamics, fluid mechanics, and mechanics of material, to predict the mechanical behaviour of soils. Together with Rock mechanics, it is the basis for solving many engineering problems in civil engineering, geophysical engineering and engineering geology. Some of the basic theories of soil mechanics are the basic description and classification of soil, effective stress, shear strength, consolidation, lateral earth pressure, bearing capacity, slope stability, and permeability. Foundations, embankments, retaining walls, earthworks and underground openings are all designed in part with theories from soil mechanics.

Characteristics of Soils

Soil is usually composed of three phases: solid, liquid, and gas. The mechanical properties of soils depend directly on the interactions of these phases with each other and with applied potentials.

The solid phase of soils contains various amounts of crystalline clay and non-clay minerals, noncrystalline clay material, organic matter, and precipitated salts. These minerals are commonly formed by atoms of elements such as oxygen, silicon, hydrogen, and aluminum, organised in various crystalline forms. These elements along with calcium, sodium, potassium, magnesium, and carbon comprise over 99% of the solid mass of soils. Although, the amount of non-clay material is greater than that of clay and organic material, the latter have a greater influence in the behaviour of soils. Solid particles are classified by size as clay, silt, sand, gravel, cobbles, or boulders.

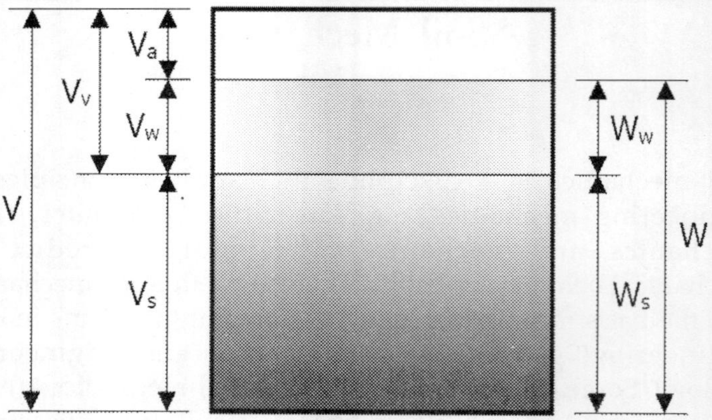

Figure 1. A phase diagram of soil indicating the weights and volumes of air, soil, water, and voids.

The liquid phase in soils is commonly composed of water containing various types and amounts of dissolved electrolytes. Organic compounds, both soluble and immiscible are present in soils from chemical spills, leaking wastes, and contaminated groundwater. The gas phase, in partially saturated soils, is usually air, although organic gases may be present in zones of high biological activity or in chemically contaminated soils. Soil mineralogy controls the size, shape, and physical and chemical properties of soil particles and thus its load-carrying ability and compressibility.

The structure of a soil is the combined effects of fabric (particle association, geometrical arrangement of particles, particle groups, and pore spaces in a soil), composition, and interparticle forces. The structure of soils is also use to account for differences between the properties of natural (structured) and remolded soils (destructured). The structure of a soil reflects all facets of the soil composition, history, present state, and environment.

Initial conditions dominate the structure of young deposits at high porosity or freshly compacted soils; whereas older soils at lower porosity reflect the post-depositional changes more. Soil, like any other engineering material, distorts when placed under a load. This distortion is of two kinds—shearing, or sliding, distortion and compression. In general, soils cannot withstand tension. In some situations the particles can be cemented together and a small amount of tension may be withstood, but not for long periods.

Particles of sands and many gravels consist overwhelmingly of silica. They can be rounded due to abrasion while being transported by wind or water, or sharp-cornered, or anything in between, and are roughly equi-dimensional.

Clay particles arise from weathering of rock crystals like feldspar, and commonly consist of alumino-silicate minerals. They generally have a flake-shape with a large surface area compared with their mass. As their mass is extremely small, their behaviour is governed by forces of electrostatic attraction and repulsion on their surfaces. These forces attract and adsorb water to their surfaces, with the thickness of the layer being affected by dissolved salts in the water.

Concept of Effective Stress

The concept of effective stress is one of Karl Terzaghi's most important contributions to soil mechanics. It is a measure of the stress on the soil skeleton (the collection of particles in contact with each other), and determines the ability of soil to resist shear stress. It cannot be measured in itself, but must be calculated from the difference between two parameters that can be measured or estimated with reasonable accuracy.

Effective stress (σ') on a plane within a soil mass is the difference between total stress (σ) and pore water pressure (u):

$$\sigma' = \sigma - u$$

Much like the concept of stress itself, the formula is a construct, for the easier visualisation of forces acting on a soil mass, especially simple analysis models for slope stability, involving a slip plane. With these models, it is important to know the total weight of the soil above (including water), and the pore water pressure within the slip plane, assuming it is acting as a confined layer.

However, the formula becomes confusing when considering the true behaviour of the soil particles under different measurable conditions, since none of the parameters are actually independent actors on the particles. Consider a grouping of round quartz sand grains, piled loosely, in a classic 'cannonball' arrangement. As can be seen, there is a contact stress where the spheres actually touch. Pile on more spheres and the contact stresses increase, to the point of causing frictional instability (dynamic friction), and perhaps failure.

The independent parameter affecting the contacts (both normal and shear) is the force of the spheres above. This can be calculated by using the overall average density of the spheres and the height of spheres above. If we then have these spheres in a beaker and add some water, they will begin to float a little depending on their density (buoyancy). With natural soil materials, the effect can be significant, as anyone who has lifted a large rock out of a lake can attest. The contact stress on the spheres decreases as the beaker is filled to the top of the spheres, but then nothing changes if more water is added.

Although the water pressure between the spheres (pore water pressure) is increasing, the effective stress remains the same, because the concept of 'total stress' includes the weight of all the water above. This is where the equation can become confusing, and the effective stress can be calculated using the buoyant density of the spheres (soil), and the height of the soil above. The concept of effective stress truly becomes interesting when dealing with non-hydrostatic pore water pressure. Under the conditions of a pore pressure gradient, the ground water flows, according to the permeability equation.

Using our spheres as a model, this is the same as injecting (or withdrawing) water between the spheres. If water is being injected, the seepage force acts to separate the spheres and reduces the effective stress. Thus, the soil mass becomes weaker. If water is being withdrawn, the spheres are forced together and the effective stress increases. Two extremes of this effect are quicksand, where the groundwater gradient and seepage force act against gravity; and the 'sandcastle effect', where the water drainage and capillary action act to strengthen the sand. As well, effective stress plays an important role in slope stability, and other geotechnical engineering and engineering geology problems, such as groundwater-related subsidence.

Total Stress

The total stress σ is equal to the overburden pressure or stress, which is made up of the weight of soil vertically above the plane, together with any forces acting on the soil surface. Total stress increases with increasing depth in proportion to the density of the overlying soil.

Pore Water Pressure

The pore water pressure u is the pressure of the water on that plane in the soil, and is most commonly calculated as the hydrostatic pressure. For stability calculations in conditions of dynamic flow (under sheet piling, beneath a dam toe, or within a slope, for instance), u must be estimated from a flow net. In the situation of a horizontal water table pore water pressure increases linearly with increasing depth below it.

Shear Strength

Shear strength in reference to soil is a term used to describe the maximum strength of soil at which point significant plastic deformation or yielding occurs due to an applied shear stress. There is no definitive "shear strength" of a soil as it depends on a number of factors affecting the soil at any given time and on the frame of reference, in particular the rate at which the shearing occurs.

Two theories are commonly used to estimate the shear strength of a soil depending on the rate of shearing as a frame of reference. These are Tresca theory for short term loading of a soil, commonly referred to as the undrained strength or the total stress condition and Mohr-Coulomb theory combined with the principle of effective stress for the long term loading of a soil, commonly referred to as the drained strength or the effective stress condition.

In modern soil mechanics, both these classical approaches (Tresca and Mohr-Coulomb) may be superseded by critical state theory or by steady state theory either of which can be considered in both undrained and drained terms and also cases involving partial drainage. The classical approaches are still in common usage; however, both are taught in undergraduate civil engineering programmes, and consequently, they are also used in practice.

Shear strength of a soil is also of importance in designing for earthquakes where the concept of the soil's steady state shear strength is used. The shear strength of soils is taught in detail in specialist masters degree programs. Such programs usually include the use of modern numerical modelling techniques such as finite element analysis coupled with a model for shear strength such as critical state soil mechanics.

Factors Controlling Shear Strength of Soils

The stress-strain relationship of soils, and therefore the shearing strength, is affected by:

- Soil composition (basic soil material): mineralogy, grain size and grain size distribution, shape of particles, pore fluid type and content, ions on grain and in pore fluid.
- State (initial): Define by the initial void ratio, effective normal stress and shear stress (stress history). State can be describe by terms such as: loose, dense, overconsolidated, normally consolidated, stiff, soft, contractive, dilative, etc.

— Structure: Refers to the arrangement of particles within the soil mass; the manner the particles are packed or distributed. Features such as layers, joints, fissures, slickensides, voids, pockets, cementation, etc, are part of the structure. Structure of soils is described by terms such as: undisturbed, disturbed, remolded, compacted, cemented; flocculent, honey-combed, single-grained; flocculated, deflocculated; stratified, layered, laminated; isotropic and anisotropic.

— Loading conditions: Effective stress path, i.e., drained, and undrained; and type of loading, i.e., magnitude, rate (static, dynamic), and time history (monotonic, cyclic)).

Undrained Strength

This term describes a type of shear strength in soil mechanics as distinct from drained strength.

Conceptually, there is no such thing as the undrained strength of a soil. It depends on a number of factors, the main ones being:

— Orientation of stresses

— Stress path

— Rate of shearing

— Volume of material (like for fissured clays or rock mass)

Undrained strength is typically defined by Tresca theory, based on Mohr's circle as:

$$\sigma_1 - \sigma_3 = 2\ S_u$$

Where:

σ_1 is the major principal stress

σ_3 is the minor principal stress

τ is the shear strength $(\sigma_1 - \sigma_3)/2$

hence, $\tau = S_u$ (or sometimes c_u), the undrained strength.

It is commonly adopted in limit equilibrium analyses where the rate of loading is very much greater than the rate at which pore water pressures, that are generated due to the action of shearing the soil, may dissipate. An example of this is rapid loading of sands during an earthquake, or the failure of a clay slope during heavy rain, and applies to most failures that occur during construction. As an implication of undrained condition, no elastic volumetric strains occur, and thus Poisson's ratio is assumed to remain 0.5 throughout shearing.

The Tresca soil model also assumes no plastic volumetric strains occur. This is of significance in more advanced analyses such as in finite element analysis. In these advanced analysis methods, soil models other than Tresca may be used to model the undrained condition including Mohr-Coulomb and critical state soil models such as the modified Cam-clay model, provided Poisson's ratio is maintained at 0.5.

One important empirical relationship used extensively by practicing engineers is the empirical SHANSEP (stress history and normalised soil engineering properties) relationship. This is based on the observation that the logarithm of the undrained shear strength Su normalised by the vertical consolidation stress svc plots linearly against the logarithm of the over consolidation ratio or OCR, that is, Su/svc=K*OCRN where K and N are constants that depend on the soil and the loading used to shear the soil. To date, no physical model has been proposed that explains this empirical observation.

Drained Strength

This term describes a type of shear strength in soil mechanics as distinct from undrained strength.

The drained strength is the strength of the soil when pore water pressures, generated during the course of shearing the soil, are able to rapidly dissipate. It also applies where no pore water exists in the soil (the soil is dry). It is commonly defined using Mohr-Coulomb theory (it was called "Coulomb's equation" by Karl von Terzaghi in 1942 combined with the principle of effective stress.

Drained strength is defined as:

$$\tau = \sigma' \tan(\phi') + c'$$

Where $\sigma' = (\sigma - u)$, known as the principle of effective stress. s is the total stress applied normal to the shear plane, and u is the pore water pressure acting on the same plane.

ϕ' = the effective angle of shearing resistance. Formerly termed 'angle of internal friction' after Coulomb friction, where the coefficient of friction μ is equal to $\tan(\phi)$, which is proportional to the normal force on a plane but independent of its area. It is now regarded to have little to do with friction, and more to do with the micro-mechanical interaction of soil particles.

It has sometimes been referred to as the "angle of repose" as a dry granular material will form a pile at this angle but no steeper. It is further described as either peak f'p, critical state ϕ'_{cv} or residual ϕ'_r. Note that ϕ'_p is only adopted in relation to Terzaghi's misunderstanding of the nature of "true" cohesion. Nowadays, critical state ϕ'_{cv} values should be prescribed.

c' = apparent cohesion. Allows the soil to possess some shear strength at no confining stress, or even under tensile stress. Commonly ascribed to temporary negative pore water pressures (suction), that dissipate over time. It may also be due to diagenetic affects caused by soil aging such as chemical bonding, cementation of grains and the effects of creep; indeed Coulomb identified that soil possessed no cohesion when newly remoulded, Heyman 1972 as these diagenetic effects had been destroyed. When shear tests are conducted on an overconsolidated or dense soil, and peak strengths are plotted on a t/s plot, it appears that cohesion exists as the y-intercept is non-zero. Some feel that this is not due to true cohesion, but is the effect of interlocking of particles.

In any case, the long term loading condition must rely on the soil properties expected to exist and contribute to the shear strength of the soil over the long term, and for these reasons it is generally not considered a reliable soil mechanical property unlike ϕ'.

Critical State Soil Mechanics

A more advanced understanding of the behaviour of soil undergoing shearing lead to the development of the critical state theory of soil mechanics. In critical state soil mechanics, a distinct shear strength is identified where the soil undergoing shear does so at a constant volume, also called the 'critical state'. Thus there are three commonly identified shear strengths for a soil undergoing shear:

— Peak strength τ_p

— Critical state or constant volume strength τ_{cv}

— Residual strength τ_r

The peak strength may occur before or at critical state, depending on the initial state of the soil particles being sheared:

— A loose soil will contract in volume on shearing, and may not develop any peak strength above critical state. In this case 'peak' strength will coincide with the critical state shear strength, once the soil has ceased contracting in volume. It may be stated that such soils do not exhibit a distinct 'peak strength'.

— A dense soil may contract slightly before granular interlock prevents further contraction (granular interlock is dependent on the shape of the grains and their initial packing arrangement). In order to continue shearing once granular interlock has occurred, the soil must dilate (expand in volume). As additional shear force is required to dilate the soil, a 'peak' strength occurs. Once this peak strength caused by dilation has been overcome through continued shearing, the resistance provided by the soil to the applied shear stress reduces (termed "strain softening"). Strain softening will continue until no further changes in volume of the soil occur on continued shearing. Peak strengths are also observed in overconsolidated clays where the natural fabric of the soil must be destroyed prior to reaching constant volume shearing. Other affects that result in peak strengths include cementation and bonding of particles.

The constant volume (or critical state) shear strength is said to be intrinsic to the soil, and independent of the initial density or packing arrangement of the soil grains. In this state the grains being sheared are said to be 'tumbling' over one another, with no significant granular interlock or sliding plane development affecting the resistance to shearing. At this point, no inherited fabric or bonding of the soil grains affects the soil strength.

The residual strength occurs for some soils where the shape of the particles that make up the soil become aligned during shearing (forming a slickenside), resulting in reduced resistance to continued shearing (further strain softening). This is particularly true for most clays that comprise plate-like minerals, but is also observed in some granular soils with more elongate shaped grains. Clays that do not have plate-like minerals (like allophanic clays) do not tend to exhibit residual strengths.

Use in practice: If one is to adopt critical state theory and take $c' = 0$; τ_p may be used, provided the level of anticipated strains are taken into account, and the effects of potential rupture or strain softening to critical state strengths are considered. For large strain deformation, the potential to form slickensided surface with a f'r should be considered (such as pile driving). The Critical State occurs at the quasi-static strain rate. It does not allow for differences in shear strength based on different strain rates. Also at the critical state, there is no particle alignment or specific soil structure.

Steady State Strength

The steady state strength is defined as the shear strength of the soil when it is at the steady state condition. The steady state condition is defined as "that state in which the mass is continuously deforming at constant volume, constant normal effective stress, constant shear stress, and constant velocity." Steve Poulos built off a hypothesis that Arthur Casagrande was formulating towards the end of his career. Steady state based soil mechanics is sometimes called "Harvard soil mechanics".

The steady state occurs only after all particle breakage if any is complete and all the particles are oriented in a statistically steady state condition and so that the shear stress needed to continue deformation at a constant velocity of deformation does not change. The steady state which can occur at any different strain rate, but has a slightly different value depending on the strain rate at which it is measured. Thus the steady state shear strength at the quasi-static strain rate would seem to correspond to the critical state shear strength.

However there is an additional difference between the two states. This is that at the steady state condition the grains align in the direction of shear, forming a special structure called the steady state structure, whereas no such oriented structure occurs for the steady state. In this sense the steady state corresponds to the "residual" condition. The steady state applies to both the drained condition and to the undrained condition. Geotechnical engineers use the steady state shear strength when designing for earthquakes.

Harvard University's soil mechanics department shut down in 1970 after Casagrande retired but even today, the concept of the steady state condition remains powerfully relevant, a reflection of the stagnation of research in soil shear. A primer on the Steady State theory can be found in a report by Poulos. Its use in earthquake engineering is described in detail in another publication by Poulos.

Consolidation Process

Consolidation is a process by which soils decrease in volume. It occurs when stress is applied to a soil that causes the soil particles to pack together more tightly, therefore reducing its bulk volume. When this occurs in a soil that is saturated with water, water will be squeezed out of the soil. The magnitude of consolidation can be predicted by many different methods.

In the Classical Method, developed by Karl von Terzaghi, soils are tested with an oedometer test to determine their compression index. This can be used to predict the amount of consolidation.

When stress is removed from a consolidated soil, the soil will rebound, regaining some of the volume it had lost in the consolidation process. If the stress is reapplied, the soil will consolidate again along a recompression curve, defined by the recompression index. The soil which had its load removed is considered to be overconsolidated. This is the case for soils which have previously had glaciers on them.

The highest stress that it has been subjected to is termed the preconsolidation stress. The over consolidation ratio or OCR is defined as the highest stress experienced divided by the current stress. A soil which is currently experiencing its highest stress is said to be normally consolidated and to have an OCR of one. A soil could be considered underconsolidated immediately after a new load is applied but before the excess pore water pressure has had time to dissipate.

Analysis of Consolidation

Spring Analogy

The process of consolidation is often explained with an idealised system composed of a spring, a container with a hole in its cover, and water. In this system, the spring represents the compressibility or the structure itself of the soil, and the water which fills the container represents the pore water in the soil.

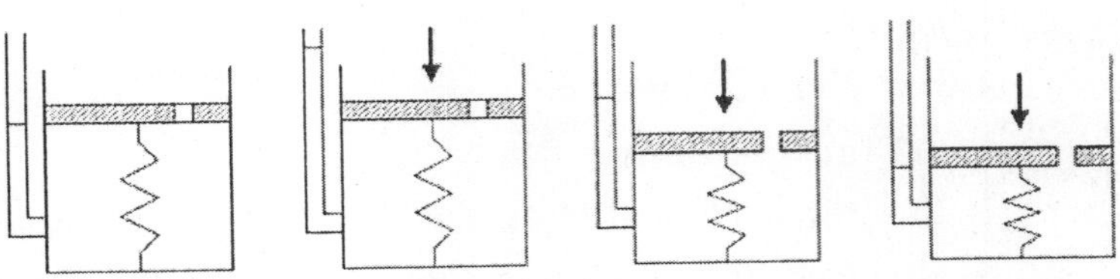

(1) (2) (3) (4)

Figure 2.

- The container is completely filled with water, and the hole is closed. (Fully saturated soil)
- A load is applied onto the cover, while the hole is still unopened. At this stage, only the water resists the applied load. (Development of excessive pore water pressure)
- As soon as the hole is opened, water starts to drain out through the hole and the spring shortens. (Drainage of excessive pore water)
- After some time, the drainage of water no longer occurs. Now, the spring alone resists the applied load. (Full dissipation of excessive pore water pressure. End of consolidation)

Primary Consolidation

This method assumes consolidation occurs in only one-dimension. Laboratory data is used to construct a plot of strain or void ratio verses effective stress where the effective stress axis is on a logarithmic scale. The plot's slope is the compression index or recompession index. The equation for consolidation settlement of a normally consolidated soil can then be determined to be:

$$\delta_c = \frac{C_c}{1+e_0} H \log\left(\frac{\sigma'_{zf}}{\sigma'_{z0}}\right)$$

where

δ_c is the settlement due to consolidation.

C_c is the compression index.

e_0 is the initial void ratio.

H is the height of the soil.

σ_{zf} is the final vertical stress.

σ_{z0} is the initial vertical stress.

C_c can be replaced by C_r (the recompession index) for use in overconsolidated soils where the final effective stress is less than the preconsolidation stress. When the final effective stress is greater than the preconsolidation stress, the two equations must be used in combination to model both the recompression portion and the virgin compression portion of the consolidation process, as follows:

$$\delta_c = \frac{C_r}{1+e_0} H \log\left(\frac{\sigma'_{zf}}{\sigma'_{z0}}\right) + \frac{C_c}{1+e_0} H \log\left(\frac{\sigma'_{zf}}{\sigma'_{zc}}\right)$$

where σ_{zc} is the preconsolidation stress of the soil.

Secondary Consolidation

Secondary consolidation is the compression of soil that takes place after primary consolidation. Secondary consolidation is caused by creep, viscous behaviour of the clay-water system, compression of organic matter, and other processes. In sand, settlement caused by secondary compression is negligible, but in peat, it is very significant.

Secondary consolidation is given by the formula:

$$S_s = \frac{H_0}{1+e_0} C_a \log\left(\frac{t}{t_{90}}\right)$$

Where

H_0 is the height of the consolidating medium

e_0 is the initial void ratio

C_a is the secondary compression index

Time Dependency

The time for consolidation to occur can be predicted. Sometimes consolidation can take years. This is especially true in saturated clays because their hydraulic conductivity is extremely low, and this causes the water to take an exceptionally long time to drain out of the soil. While drainage is occurring, the pore water pressure is greater than normal because it is carrying part of the applied stress (as opposed to the soil particles).

Lateral Earth Pressure

Lateral earth pressure is the pressure that soil exerts in the horizontal plane. The common applications of lateral earth pressure theory are for the design of ground engineering structures such as retaining walls, basements, tunnels, and to determine the friction on the sides of deep foundations.

To describe the pressure a soil will exert, a lateral earth pressure coefficient, K, is used. K is the ratio of lateral (horizontal) pressure to vertical pressure ($K = \sigma_h'/\sigma_v'$). Thus horizontal earth pressure is assumed to be directly proportional to the vertical pressure at any given point in the soil profile. K can depend on the soil properties and the stress history of the soil. Lateral earth pressure coefficients are broken up into three categories: at-rest, active, and passive.

The pressure coefficient used in geotechnical engineering analyses depends on the characteristics of its application. There are many theories for predicting lateral earth pressure; some are empirically based, and some are analytically derived.

Rest Lateral Earth Pressure

At rest lateral earth pressure, represented as K_0, is the in-situ horizontal pressure. It can be measured directly by a dilatometer test (DMT) or a borehole pressuremeter test (PMT). As these are rather expensive tests, empirical relations have been created in order to predict at rest pressure with less involved soil testing, and relate to the angle of shearing resistance.

Active and Passive State of Pressure

The active state occurs when a soil mass is allowed to relax or move outward to the point of reaching the limiting strength of the soil; that is, the soil is at the failure condition in extension. Thus it is the minimum lateral soil pressure that may be exerted. Conversely, the passive state occurs when a soil mass is externally forced to the limiting strength of the soil in compression. It is the maximum lateral soil pressure that may be exerted. Thus active and passive pressures define the minimum and maximum possible pressures respectively that may be exerted in a horizontal plane.

Bearing Capacity

In geotechnical engineering, bearing capacity is the capacity of soil to support the loads applied to the ground. The bearing capacity of soil is the maximum average contact pressure between the foundation and the soil which should not produce shear failure in the soil. Ultimate bearing capacity is the theoretical maximum pressure which can be supported without failure; while allowable bearing capacity is the ultimate bearing capacity divided by a factor of safety. Sometimes, on soft soil sites, large settlements may occur under loaded foundations without actual shear failure occurring; in such cases, the allowable bearing capacity is based on the maximum allowable settlement.

There are three modes of failure that limit bearing capacity: general shear failure, local shear failure, and punching shear failure.

Slope Stability

The field of slope stability encompasses the analysis of static and dynamic stability of slopes of earth and rock-fill dams, slopes of other types of embankments, excavated slopes, and natural slopes in soil and soft rock.

Earthen slopes can develop a cut-spherical weakness zone. The probability of this happening can be calculated in advance using a simple 2-D circular analysis package. A primary difficulty with analysis is locating the most-probable slip plane for any given situation. Many landslides have only been analysed after the fact. Slope stability issues can be seen with almost any walk down a ravine in an urban setting.

Analysis Methods

If the forces available to resist movement are greater than the forces driving movement, the slope is considered stable. A factor of safety is calculated by dividing the forces resisting movement by the forces driving movement. In earthquake-prone areas, the analysis is typically run for static conditions and pseudo-static conditions, where the seismic forces from an earthquake are assumed to add static loads to the analysis.

Method of Slices

The method of slices is a method for analysing the stability of a slope in two dimensions. The sliding mass above the failure surface is divided into a number of slices. The forces acting on each slice are obtained by considering the mechanical equilibrium for the slices.

Bishop's Method

The Modified (or Simplified) Bishop's Method is a method for calculating the stability of slopes. It is an extension of the Method of Slices. By making some simplifying assumptions, the problem becomes statically determinate and suitable for hand calculations:

— forces on the sides of each slice are horisontal

The method has been shown to produce factor of safety values within a few percent of the "correct" values.

Lorimer's Method

Lorimer's Method is a technique for evaluating slope stability in cohesive soils. It differs from Bishop's Method in that it uses a clothoid slip surface in place of a circle. This mode of failure was determined experimentally to account for effects of particle cementation. The method was developed in the 1930s by Gerhardt Lorimer, a student of geotechnical pioneer Karl Terzaghi.

Seepage

Seepage is the flow of a fluid through soil pores. After measuring or estimating the intrinsic permeability (K_i), one can calculate the hydraulic conductivity (K) of a soil, and the rate of seepage can be estimated. K has the units m/s and is the average velocity of water passing through a porous medium under a unit hydraulic gradient. It is the proportionality constant between average velocity and hydraulic gradient in Darcy's Law. In most natural and engineering situations the hydraulic gradient is less than one, so the value of K for a soil generally represents the maximum likely velocity of seepage.

A typical value of hydraulic conductivity for natural sands is around 1x10-3m/s, while K for clays is similar to that of concrete. The quantity of seepage under dams and sheet piling can be estimated using the graphical construction known as a flownet. When the seepage velocity is great enough, erosion can occur because of the frictional drag exerted on the soil particles. Vertically upwards seepage is a source of danger on the downstream side of sheet piling and beneath the toe of a dam or levee.

Erosion of the soil, known as "piping", can lead to failure of the structure and to sinkhole formation. Seeping water removes soil, starting from the exit point of the seepage, and erosion advances upgradient. The term sand boil is used to describe the appearance of the discharging end of an active soil pipe. Seepage in an upward direction reduces the effective stress within soil. In cases where the hydraulic gradient is equal to or greater than the critical gradient, effective stress is reduced to zero. When this occurs in a non-cohesive soil, a "quick" condition is reached and the soil becomes a heavy fluid. Quicksand was so named because the soil particles move around and appear to be 'alive'.

References

Dooley, Alan., Sandboils 101: Corps has experience dealing with common flood danger. US Army Corps of Engineers. Retrieved on 2006-08-29, June, 2006.

Holtz, R.D, and Kovacs, W.D., *An Introduction to Geotechnical Engineering*, Prentice-Hall, Inc., 1981, page 448.

Jones, J. A. A., "Soil piping and stream channel initiation", *Water Resources Research* 7 (3): 602 - 610, 1976.

Mitchell, J.K., *Fundamentals of Soil Behaviour*, John Wiley & Sons, Inc., 1993.

Poulos, S. J., *Advance Dam Engineering for Design, Construction, and Rehabilitation: Liquefaction Related Phenomena*, Ed. Jansen, R.B, Van Nostrand Reinhold, 1989, pages 292-297.

Terzaghi, K., Peck, R.B., and Mesri, G., *Soil Mechanics in Engineering Practice*, Third Edition, John Wiley & Sons, Inc. Article 18, 1996, page 135.

3

Soil Testing

Soil testing as a management tool is greatly under utilized. The most recent data for lawns and gardens are from 1987. The frequency of soil testing is the highest in the Southern States where it is one test per about 200 people per year. The per capita rate of soil testing is very low in the Southwestern States. California has a rate of one test per 2,091 people per year while Arizona has a rate of one test per 958 people per year. Ideally, every site should be tested every few years. The frequency of needed soil testing depends upon the amount of irrigation, the quality of irrigation water, the use of nutrients and amendments and the initial soil properties. Testing could help to prevent and solve many problems.

Soil

Soil is formed from the parent minerals contained in rocks. Through the influence of climate (rain, wind, heating, freezing etc.) and organisms the rocks weather. Simple plants like lichens and microorganisms use the minerals released in the weathering process and continue with the formation of soil. As organisms grow and die, organic matter accumulates which interacts with the mineral particles. Eventually, a horizon or profile of developed soil is generated which is called a topsoil.

Rain leaches the soluble minerals into the deeper soil profiles. The topsoil profile is called an "A horizon" while the soil profile which receives the minerals that are moved into the soil by water is called a "B horizon" or subsoil. Below the B horizon is the "C horizon" or the unweathered rock. The properties of the topsoil depend mainly upon native vegetation and upon the amount of rain. These are a function of climate.

Weathering of rocks releases salts which had previously been encapsulated, usually as part of the structure of the rock. In the desert and semiarid zones of the Southwest, the salt content of the soil is very high due to little leaching of the soils. The salts also impart an alkaline condition to the soil. Native plant species are desert shrubs which are tolerant of the local conditions.

Due to poor plant growth, the accumulation of organic matter is generally low and the soil has poor physical properties. The soils are light coloured. The Great Plains has more rainfall than do the Western deserts; the salinity or salt level is decreased, but there is not an excessive amount of mineral leaching. The topsoil supports the growth of grasses. Organic matter accumulates and the soils have good tilth, are fertile and have a dark brown or black appearance due to the accumulation of organic matter.

As the amount of rain increases such as in the eastern regions of the country, the nutrient content of the topsoil is lower. The minerals have been leached into the deeper soil profiles. The soil is a gray-brown. The native species are broadleaf deciduous trees in high rainfall locations and the trees are needle-leaf trees in higher rainfall areas. Organic matter decomposes forming organic acids making acidic conditions. The acidity dissolves the nutrients which are leached into the groundwater by the rains. Tropical soils near the surface are red because of the extremely low level of organic matter which unmask the presence of iron oxides which are very prevalent in soils.

The fertility is extremely low and mostly what is available comes from the recycling of the nutrients from decaying vegetation and parasitic growth from plants growing on host plants. Weathering is rapid in the hot, humid conditions which releases some nutrients from the rocks. Variations of the above conditions exist. Former marine sediments can be exposed in the normally alkaline west containing deposits of sulfur or iron sulfide which were formed from sulfate ions in the ocean.

When exposed to the air, sulfuric acid is produced by oxidation leading to acidic soils. Also earth slides, erosion or grading can expose alkaline deposits from the "B" horizon in areas which are normally acidic. Soils with the best physical properties exist with the highest level of soil organic matter. This occurs in the areas with moderate rain. The extremes of too little water and of too much rain decrease soil organic matter with a reduction in the tilth of the soil.

Changes in Soil Properties

The application of soil amendments and fertilizers can increase directly or indirectly the level of soil organic matter, increase the fertility of the soil and change the salt level and pH of the soil. Acidifying fertilizers such as ammonium sulfate and soil sulfur at high rates can make soil too acidic in normally alkaline conditions. Alkaline forming fertilizers such as calcium nitrate and potassium nitrate or the addition of excessive amounts of limestone can also cause growth inhibition. It is surprising for most people to learn that plants have growth optimum conditions for nutrients; too much can be as bad as too little.

When too much fertilizer is applied which is common for many sites, the rate of plant growth is decreased. Part of the inhibition is an induced deficiency of another nutrient caused by competition. Additionally, excessive nutrients increases the salt level of the soil which interferes with the moisture absorption by most plants. In some cases, the soils are very resistant to change. This can be the situation for the arid and semiarid climates of the Southwest where limestone is present-it is extremely difficult to acidify the soil.

Plant growth problems exist in plants species adapted to acidic soils when they are grown under alkaline conditions.

Iron deficiency recognized by yellow leaves with narrow green veins is caused by the limestone and bicarbonates. In extreme cases the yellow leaves are very small; totally white leaves are easily burned from saline conditions. Special iron products are available to correct these problems. The iron deficiency condition is so common in some locals that many people believe that some species are normally yellow when they are not.

Soil Testing Management

Soil Acidity

If the soil is too acidic, aluminum is dissolved causing a specific ion toxicity. The plant growth is stunted and the leaf coloration is sometimes deep green. If the soil is too alkaline, some plant nutrients are unavailable causing a mineral deficiency. The source of these two problems can be the use of too much plant fertilizers of the incorrect type. A soil acidity (pH test) is required to know the soil acidity status. Managing alteration in soil acidity with choice of nitrogen fertilizers—Fertilizer products are not interchangeable.

Each product has a particular advantage and benefit over other materials. Use of the incorrect product will exacerbate problems while the correct on will enhance growth. For instance with nitrogen products, ammonium sulfate (21-0-0) will acidify the soil; ammonium nitrate (34-0-0) will be pH neutral if not over applied; calcium nitrate (15.5-0-0) will slightly increase the soil pH; urea (46-0-0) needs to be hydrolysed before it is available. Nitrate nitrogen will supply soil oxygen. Ammonium nitrogen consumes oxygen when it is nitrified to nitrate.

Slow-release nitrogen materials also have certain benefits. Ureaformaldehyde (38-0-0) release nitrogen according to temperature and biological activities. IBDU releases in the presence of moisture and acidity, not according to plant growth. If coated products are broken, they become rapid-release.

Presence of Limestone

If limestone (calcium carbonate or chalk) is present, acidloving plants become iron deficient unless corrective measures are taken.

Lime Requirement

In areas of high rainfall, there are inadequate levels of potassium, calcium and/or magnesium due to the acidic soil. Tests for the required level of limestone or dolomite needed to raise the soil pH to a safe level are essential.

Excess Salts in the Soil

The term used by laboratories is salinity. If salts have excessively accumulated in the soil, many plants are unable to use the moisture in the soil and may have toxicity from sodium and/or chloride. A salinity test is required to determine if this is a problem. The salinity can be controlled by leaching unless soils have drainage problems. A soil high in salinity is called "saline."

Excessive Sodium

Excessive sodium or a "sodic" soil most often has an elevated pH level. Soils high in pH values are suspect. Sodium can cause toxicity, but the more likely problem is soil compaction and poor drainage caused by the reaction of sodium on the clay.

Gypsum Requirement

Excessive sodium can be corrected with the addition of gypsum. Another cause of high pH values is the presence of bicarbonates. Gypsum is also used to precipitate the excessive bicarbonates and lower the very high soil pH values. A laboratory test shows how much gypsum is needed.

Fertility

Most plants require at least 16 nutrients. Three nutrients are supplied by the water and by the air (oxygen, hydrogen and carbon). Thirteen are mineral nutrients. If any one is too low, the plants will not grow. In some cases, too much fertilizers have been applied causing an adverse reaction. Too much phosphorus, for instance, inhibits the plant uptake of iron, manganese, zinc and copper causing induced deficiencies. The best method to determine if a problem is caused by a true deficiency or is an induced deficiency is soil testing. Soil analysis is used to assess the nutrient levels of the soil. Plant tissue testing is also used to ascertain which nutrients have reduced availability in the soil.

Toxicity

Soils may contain toxic metals. They either exist in the soil naturally or have been introduced as contaminants in amendments. Mined minerals and waste products are the frequent contaminant sources. These elements prevent plant growth. If a vegetable garden is to be grown and if the presence of heavy metals is suspected, the soil should be tested as a precaution for human poisoning. Lead can be present in urban soil at levels which do not injure plants but can accumulate in produce at levels which may harm humans. Excess levels of selenium and molybdenum are problems with wild life or cattle. Other common toxic elements are aluminum, cadmium, chromium, nickel, arsenic, silver, and vanadium.

Soil Compaction

Excessive compaction impedes root growth, impairs water penetration and reduces soil aeration. Reduced aeration hinders the absorption of nutrients. In addition, slow water penetration exasperates the problem. Soil compaction can be measured and corrected with soil conditioners. Their need can be detected with soil testing.

Preparation of Soil Sample

Ideally, a soil should be tested without disturbing or altering it chemically or mechanically in the process of sample preparation. This would require testing in situ, which is not technically feasible today. For the convenience of handling and to provide a homogenous mix for subsampling, soil samples are usually dried and pulverized. Subsamples of the dry, pulverized soils are either weighed or measured by volume. Galvanized containers, cast iron mortars, rubber stoppers, brass screens and a variety of other tools can contribute to contamination with iron, zinc and other micronutrients, and should not be used.

Sample handling before analysis can affect soil test results. It has been shown that drying can result in increased release of exchangeable potassium (K) in many soils and in fixation in others. The fixation tends to occur in recently fertilized soils at higher test levels. The extent of reversion on rewetting varies among soils and is seldom, if ever, complete. Increased temperature can also increase the exchangeable K levels. Dowdy and Hutcheson found that illite was the source of K release on drying and that fixation could be attributed to vermiculite or montmorillonite.

Early studies in Iowa showed that the results from field-moist samples were better correlated with the potassium uptake by plants than the results from air-dried soils. Higher correlations with field-moist samples were also found in the regional K studies in the late 1950s and early 1960s. The K release on drying and the reversion on rewetting can be controlled with organic additives, but this procedure has not been evaluated in practical soil testing.

Drying and method of drying may also affect the results of the tests for mineralizable nitrogen, phosphorous, sulfur, zinc and perhaps other micronutrients, but the correlations between the test results and the uptake of nutrients by plants have not been shown to be significantly affected by drying. Primarily because of the effect of drying on potassium results, a method of testing undried soil samples was developed and put into use in the Iowa State University Soil Testing Laboratory until 1990. Because of the difficulties of analyzing moist samples and because most correlation and calibration studies have been done on air-dried soils, the undried soil analysis method has not been adopted widely. The traditional method of preparing dry samples is presented here.

Procedure for Handling Dry Soil

Traditionally, most soil analysts have considered dry soil as the convenient state from which to start chemical tests. Because soil samples are received in a wide range of physical conditions, a common denominator in preparation is required to alleviate these problems and expedite processing.

Drying

Moist, well-mixed samples may be transferred to paper bags, cardboard boxes or aluminum trays of convenient size. The open sample container is then placed in a drying rack or cabinet equipped with exhaust fans to expedite air movement and moisture loss. If heat is necessary, the temperature of the cabinet should not exceed 40°C (104°F). This is especially critical for potassium analysis, which can be significantly influenced by drying temperatures.

If nitrate analyses are involved, the soil should be dried or frozen within 12 hours of sampling. Such samples can be dried by spreading them out on a clean paper or cloth and blow drying them with a fan. Where sample volume is not adequate to justify artificial drying, samples may be spread on clean surfaces, such as paper plates. Initial crushing of soil clods will decrease the time required for drying at room temperatures. Microwave drying is a relatively rapid method to dry a few soil samples. For moisture determination, the method worked well. However, microwave drying appears to change many nutrient analyses as compared to air-drying, and is not recommended.

Crushing and Sieving Soil Sample

The nature of analyses to be conducted, plus presence of rocks or limestone concretions, dictate initial steps to crushing. Crush samples designated for mechanical analyses with a wooden rolling pin after removing all stony material from the soil. Crush other samples with a flail-type grinder, a power-driven mortar and pestle, or some other crusher which is designed to minimize contamination through carryover from one sample to another.

If micronutrient analyses are to be performed, it is essential that all surfaces coming into contact with the soil be stainless steel, plastic or wooden, preferably in the order listed. Samples should be crushed until a major portion of the sample will pass a U.S. No. 10 (2 mm opening) sieve. Crushing to pass a finer mesh sieve may be desirable for analysis utilizing less than one gram of soil.

Soil Scoop

One step in soil testing is measuring an amount of soil from which the analysis will be performed. Controversy exists concerning the merits of using a measured volume of soil versus using a predetermined weight. Where a volume measurement of sample is taken, the results may be expressed either on a direct volume basis or on a corrected weight/volume basis. Because differences existed in the test values of soil samples exchanged among member laboratories using the "same" methods, an NCR-13 committee was established to evaluate the impact of sample measurement in 1965. It became apparent early in the study that each of the different measurement techniques had its strong and weak points, and adoption of a standard technique would be a compromise between variations in soil bulk density and the fact that plants grow in a volume of soil, not a weight of soil.

Weight and Volume Measurement

Weight measurement offers the advantage of precision in determining sample size. It has the disadvantage of requiring more time, greater initial expense for weighing equipment and a larger work space area. In addition, an estimation of soil bulk density or measurement to correct for soil variation is required. Most research soil analyses are made on weighed samples.

Volume measurement is the technique most commonly used in soil testing. It has the advantage of being rapid, low cost, requires little space and integrates bulk density into the sample measurement. It has the disadvantage of reduced precision between replicate volume samples. Test results from volume sample measurements may be expressed in two ways:

(a) some volume unit, or

(b) a weight/volume basis.

Reporting soil analysis on a strictly volume basis, e.g., mg/ dm^3, is suggested by Mehlich. The weight-volume basis of reporting has been the conventional reporting method under the U.S. measurement system. The concept of the weight-volume basis is that a scoop of an appropriate size will hold a unit amount of the typical soil weighing 2 million pounds per acre to a depth of 62/3 inches. As one changes to the metric system these convenient, conventional reporting units may have conversion disadvantages.

Historical Scoop

An initial study in 1965 showed that measurement scoops of two sizes were predominant in use among the 13 member states. These sizes were approximately 0.85 cc and 1.0 cc per gram of "typical soil." Scoop construction varied greatly, consisting of modified kitchen measuring spoons, calibrated copper tubing caps and machined brass scoops. The basis for calibrating scoops was undocumented and vague. Conditions of wear and shape varied greatly, contributing further to disarray in soil measurement.

A search into the heritage of the two scoop sizes showed that the 1.0 cc scoop was introduced in the late 1950s in Illinois during a modernization of the soil testing program, and the 0.85 cc size scoop was used by developers of early soil testing methods. The magnitude of the variations in test results was not as serious as variations in scoop size due to the small degree of dissociation of nutrient forms measured by available tests.

Development of NCR-13 Scoop

In 1967, the NCR-13 representatives using volume measurement elected to adopt a standard size, stainless-steel scoop to minimize the following problems: contamination of soil samples, wear of the scoop and variation among laboratories. The committee members also decided to continue using the weight-volume basis of reporting soil test levels since soil testing with these terms has appreciable farm acceptance in the North Central Region.

The usual layman term is "pounds per acre," and conversion to the metric system can be ppm, designated with an asterisk to indicate scoop measurement. Studies show that the 0.85 cc size scoop approximates a 1 gram measure of typical soil. This is an empirical conclusion arrived at from the observation of several hundred volume/weight measurements on a wide range of soils.

The typical soil is defined as a medial silt-loam texture with 2.5 percent organic matter crushed to pass a 10-mesh screen. Bulk density of crushed, typical soil approximates 1.18, which compared with 1.32 for undisturbed soil. Experience with the 0.85 cc scoop shows that soil test results on a sample measured with such a scoop, when compared to a weighed sample analysis, differ by a factor equal to the difference in bulk density of the soil samples.

Procedure for Using Soil Scoop

Suggested procedure for using a soil scoop to measure soil is as follows:

1. Stir the pulverized and screened soil sample with a spatula to loosen soil prior to measuring.
2. Dip into the center of the soil sample with the scoop, filling it heaping full.
3. Hold the scoop firmly. Tap the handle three times with a spatula from a distance of 2 to 3 inches.
4. Hold the spatula blade perpendicular to the top of the scoop and strike off excess soil.
5. Empty the scoop into an extraction vessel for the soil test.
6. Calculate the analytical result using the scoop size as the assumed weight of soil and report soil test value in units of pounds per acre.

Laboratory Factors and Soil Extraction

One of the responsibilities of the NCR-13 committee is to standardize soil testing procedures for the North Central Region. These procedures include the chemical methods for extracting plant nutrients from the soil and their subsequent analysis by means that are accurate and free from interferences. However, sometimes overlooked are the details of exactly how those extractions are performed in a laboratory that is processing large numbers of samples under time pressure. This includes factors such as the means of shaking, rate of reciprocation, type of extraction vessel, extraction time and laboratory temperature.

Although these factors are sometimes not given a lot of attention, they can have a significant impact on the test results. As laboratories work to improve the accuracy and reproducibility of their testing, it is important that they examine these factors and the impact they may have on test results. The more agreement that exists in this area, the better chance there will be for uniformity of test results across laboratories. In the pursuit of consistency among laboratories, some of the following points should be considered when examining the techniques used in extracting plant nutrients by the recommended standard procedures.

Extraction Vessel Shape

Studies by Grava have shown that Erlenmeyer flasks are preferred to ensure adequate mixing of the extracting solution and soil. The size of the flasks used in a particular test is determined by the volume of the extracting solution needed. The guideline to follow is that the flask should be about one-fourth full for best agitation. Wheaton bottles or similar straight sided bottles are discouraged due to the risk of variable and inadequate mixing.

Shaking and Stirring

Stirring has sometimes been used in place of shaking for mixing soil with an extracting solution. Grava found this is acceptable if a stirring rate of 500 rpm is used. Agboola and Omueti found that stirring extracted more P than shaking with the Bray P-1 test and felt shaking to be a better method for careful work.

Studies by Munter did not show large differences in extractability of nutrients with variable shaking rates over the range of 160 to 260 epm (excursions per minute). A study by Stone reported lower sodium bicarbonate extractable phosphorus (P) with rates of 140 epm. Grava recommended a shaking rate of 160 to 240 epm when using Erlenmeyer flasks. This recommendation was also endorsed by Munter.

Extraction Time

Each test has a recommended extraction (shaking/ reflux) time that should be strictly followed. A number of studies have shown that for some tests, an extraction time other than the recommended time can have significant effects on the amount of nutrient extracted. For example, with the DTPA test, it has been found that shaking 15 minutes beyond the 2-hour recommended shaking time can result in a significant increase in the amount of micronutrients extracted for some soils. As demonstrated by Grava, it is inadvisable for laboratories to cut corners by reducing shaking times, especially for P and potassium (K).

A study by Agboola and Omueti paralleled the results of Bray and Kurtz showing that Bray P-1 reached an equilibrium after 5 minutes, declined at 10 minutes of shaking and again increased at 20 minutes. These variations point out the importance of consistency in extraction time. In addition, McGeehan et al. have found that reflux and cooling times can strongly affect the results of the hot water extractable boron (B) soil test.

They found that increased reflux time resulted in higher extractable B values, while increased cooling time decreased extractable B values. This points out the importance of following the recommended 5-minute reflux time and to carefully standardize the cooling time for this test. McGeehan et al. recommend a 10-minute cooling period.

Laboratory Temperature

The temperature of the laboratory is an often overlooked factor in laboratory analysis. There are several areas where this can have an impact. One area is in pH measurements. If pH reference buffers are not the same temperature as the samples or if the measurements are not temperature compensated, the error can be as much as 0.05 pH units for every 4°C change in temperature. It is recommended that all solutions be at ambient laboratory temperature. Don't use reference buffers or calibration standards still cold from the refrigerator or distilled water directly from a cold tap. Make sure the pH meter is properly adjusted for temperature.

The extractability of nutrients has also been found to vary with changes in the ambient laboratory temperature. Stone reported that the temperature of extraction was an important factor in sodium bicarbonate extractable P, the extractable P increasing with temperature over the range of 14 to 33°C. Olsen found that sodium bicarbonate extractable P increased 0.43 ppm P with each 1°C increase in ambient temperature between 20 to 30°C. Munter also reported strong temperature effects on Bray 1 extractable P, with P levels increasing as much as 126 percent with a temperature increase from 24 to 35°C. For this reason, Munter strongly recommended that laboratory temperatures be maintained between 24 to 27°C for routine extraction.

Soil pH

Soil pH is a measure of hydronium ion activity in a soil suspension. This property influences many aspects of crop production and soil chemistry, including availability of nutrients and toxic substances, activity and diversity of microbial populations, and activity of certain pesticides. Soil pH is defined as the negative logarithm (base 10) of the H^+ activity (moles per liter) in the soil solution.

As the activity of H^+ in the soil solution increases, the soil pH value decreases. Soils with pH values below 7 are referred to as "acid"; pH values above 7 are referred to as "alkaline";

soils at pH 7 are referred to as "neutral." In most soils, the soil pH is buffered by several components of the solid phase, including hydroxy aluminum monomers and polymers, soil organic matter, and undissolved carbonates in soils. Lime requirement tests, which generate recommendations for effecting relatively long-term changes in soil pH, are designed to account for soil buffering capacity.

Determination of Soil pH

Soil pH is usually measured potentiometrically in a slurry using an electronic pH meter. A H^+ sensitive electrode and a reference electrode are also used. Combination electrodes that contain the H^+ sensitive electrode and the reference electrode as one unit can be used if the combination electrode is robust enough to withstand continued wear from the soil slurry over time. Instructions for the correct pH meter operation are provided by the manufacturer.

Several precautions should be taken when measuring pH of a soil/liquid slurry. Electrodes should be checked and maintained frequently to prevent surface residue buildup, which may affect the measurement. Rinsing between each soil sample, however, is not usually necessary. Electrodes should be protected to prevent insertion to the very bottom of the slurry-containing vessel. If this is not done, abrasion of the sensing surfaces will occur, decreasing the life of the electrode and leading to inaccurate pH readings.

All meters should be calibrated routinely at two points with buffer solutions of known pH before measuring the pH of a soil sample. One point of calibration should be at pH 7.0, while the other point should be chosen based on the range of soil pH normally encountered by the laboratory. A laboratory testing mainly acid soils should calibrate across the acid range (second point at pH 4, for example), while a laboratory testing mainly alkaline soils should calibrate across the alkaline range (second point at pH 9 or 10).

Reference and/or combination electrodes for measuring soil pH should be chosen carefully because flow rates at the liquid junction can affect the accuracy of soil pH readings.

Laboratories should use a set of reference soil samples of known pH to evaluate the performance of electrodes. Such samples should be stored and handled under carefully controlled conditions to prevent changes in soil properties over time. The reference soil pH of these samples should be determined using the average reading of several meters over several days. Electrodes that have been calibrated with clear buffer solutions, but fail to produce pH readings of the reference soils consistent with established values, should be discarded. However, the properties of the reference soils must not have changed.

Soil pH is normally measured in a soil/water slurry. The presence of soluble salts in a soil sample will affect pH. For that reason, some analysts prefer to measure pH in a mixture of soil and 0.01 M $CaCl_2$. The excess salt in this solution masks the effects of differential soluble salt concentrations in individual samples. Below are procedures for each method.

Equipment and Reagents

1. NCR-13, standard 5 g soil scoop
2. pH meter with appropriate electrode(s)
3. Pipettes
4. 1 oz. paper cups or equivalent
5. Distilled or deionized water
6. 0.01 or 1.0 M $CaCl_2$
7. Appropriate buffer solutions for calibrating the pH meter

Procedure

1. Calibrate the pH meter over the appropriate range using the manufacturer's instructions.
2. Use the scoop to measure a 5 g soil sample into a paper cup.
3. Add 5 mL distilled or deionized water to the sample.
4. Stir vigorously for 5 seconds and let stand for 10 minutes.

5. Place electrodes in the slurry, swirl carefully and read the pH immediately. Ensure that the electrode tips are in the swirled slurry and not in the overlying solution.
6. For the $CaCl_2$ measurement, add one drop of 1.0 M $CaCl_2$ solution to the previous sample, or prepare a sample as in Steps 2 and 3, using 0.01 M $CaCl_2$ instead of water.

Lime Requirement Determination

The SMP buffer method described below was designed for soils with large lime requirements and significant reserves of exchangeable aluminum. The procedure may be inaccurate on low lime requirement soils (<2 T/A), soils with organic matter contents greater than 10 percent, sandy soils or soils with a predominance of kaolinite and hydroxy oxides of aluminum and iron in their clay fractions. For most North Central Region soils, however, the SMP method appears to yield satisfactory results. For more precise determination of very low lime requirements, the double-buffer modification of the SMP procedure may prove useful.

Equipment and Reagents

1. Equipment needed for pH determination
2. Mechanical shaker
3. SMP buffer solution

SMP Buffer Solution

1. Weigh into an 18 L bottle:
   a. 32.4 g paranitrophenol1;
   b. 54.0 g potassium chromate;
   c. 955.8 g calcium chloride dihydrate.
2. Add 9 L distilled or deionized water, shaking vigorously during addition.
3. Weigh 36.0 g calcium acetate into a separate container and dissolve in 5 L of distilled or deionized water.

4. Combine solutions 2 and 3, shaking during mixing and every 15 to 20 minutes for 2 to 3 hours.
5. Add 45 mL triethanolamine, shaking during addition and periodically thereafter until completely dissolved (may take up to 8 hours). A magnetic stirrer can be use as an alternative to periodic shaking.
6. Dilute to 18 L with distilled or deionized water, adjust to pH 7.50 using 15 percent NaOH, and filter. To minimize air bubbles, avoid excessive agitation of the solution after pH adjustment.
7. Store in a container with the air inlet protected by drierite and ascarite to prevent contamination by water vapour and carbon dioxide.

Measurement of Soil-Buffer pH

1. Add 10 mL of SMP buffer solution to the soil/ water slurry saved from the pH determination.
2. Place in a mechanical shaker, close tightly, shake at 250 excursions per minute for 10 minutes, and let stand for 30 minutes.
3. Swirl and read the pH. Read to the nearest 0.01 pH unit, particularly if using the double-buffer option described below.
4. Use the resulting soil-buffer pH to determine the lime requirement.

Nitrate-Nitrogen

PPNT

Pre-plant Soil Nitrate Tests (PPNT) have been used for decades to predict crop nitrogen (N) needs in the Great Plains region. In these low-rainfall areas, nitrate carryover from the previous growing season is frequent due to low potential for nitrate loss through leaching and denitrification. Recent work in humid areas of the Midwest shows that the PPNT is a useful method of adjusting crop N recommendations for the amounts of residual nitrate in soil profiles.

In humid regions, the PPNT is likely to be most useful on medium and fine-textured soils where previous-year precipitation was normal or below normal, and where previous-year N applications exceeded crop N needs. In general, the PPNT consists of measuring nitrate-N in some portion of the crop root zone and crediting this N against the N needs of the crop to be grown.

The PPNT differs in principle from the Pre-sidedress Nitrate Test (PSNT). PPNT provides a direct crediting of soil nitrate against crop N needs, while the PSNT provides an index of N availability that is related to crop N response through test calibration data. Therefore, the PPNT does not provide a direct assessment of the amounts of N likely to be released from organic N sources during the growing season.

Procedures for using the PPNT have recently been summarized by Bundy and Meisinger. Soil sampling for the PPNT can be done in the fall (subhumid areas only) or before crops are planted in the spring. This provides more time for sampling and analysis than is usually available for the PSNT. Samples are usually taken to a minimum depth of 2 feet. Some states recommend deeper samples. Standard analytical procedures are appropriate for extraction and analysis of nitrate in the soil samples.

Samples should be dried or frozen soon after collection, and moist samples should be protected from warm temperatures. Interpretation and use of PPNT results to develop fertilizer recommendations for crops varies among states and regions, but direct crediting of nitrate-N against crop N needs is usually involved. Examples of interpretation procedures for PPNT results in humid regions are provided in Bundy and Sturgul and Schmitt and Randall.

Soil Nitrate Test for Corn

In humid regions, the nitrate-N level in the top foot of soil, measured just prior to sidedressing corn, has been shown to be related to the N-supplying capability of the soil and thus, the probability of a corn yield response to sidedress application of N fertilizer. This test, which is called the Pre-sidedress Soil Nitrate Test (PSNT), is essentially an in situ incubation method which provides an index of the N-supplying capability of a soil. It is not used to reduce recommendations by the amount of extracted nitrate-N such as is done with deep nitrate tests in the Great Plains.

The PSNT has been especially useful in helping to estimate the probability of corn response to N where manure has been applied. While this test is designed primarily to determine the probability of response to sidedress N, it can also provide some guidance for improving N rate recommendations when a response is predicted. Standard recommended procedures for extraction and analysis of the soil nitrate-N can be used. However, the timing and method of sampling for the PSNT is unique.

Samples must be taken when the corn is approximately 12 inches tall. Samples are taken to a 12 inch depth and must be dried within 24 hours. If they cannot be dried immediately, they must be frozen until they can be dried for analysis. In addition to the standard recommended procedures for extraction and analysis for soil nitrate-N, quick-test methods have been developed in Pennsylvania and Iowa in response to the need for rapid turnaround with this analysis. These kits have been shown to agree well with standard laboratory procedures and have been used successfully.

Interpretation of the soil nitrate-N levels and the method of making recommendations for sideidressing N vary from state to state. Research from the Northeast to the Midwest has indicated that the critical level for the PSNT generally ranges from 21 to 25 ppm nitrate-N. Above this level, a response to sidedress N is not expected. Below this level, recommendations can be adjusted based on the nitrate-N level from the PSNT.

Soil Extraction

The high solubility of NO_3 in water makes extraction with water possible for most soils of the North Central Region. Soils with significant anionexchange properties should be extracted with 1 or 2 molar KCl, unless Cl^- or K^+ ions interfere with the determination method. Time of extraction varies from 5 to 30 minutes for various states. Kelly and Brown found that shaking the sample for 5 minutes gave similar results to shaking for 8 hours. Oien and Selmer-Olsen found a 2-minute shaking time sufficient to extract nitrates.

The extraction of NO_3^- is relatively simple; however, many methods of determination exist. Errors can arise in each of these methods due to interferences, biological transformations, poor lab technique and many other sources. Variations due to method of determination are of relatively minor importance, though, compared to those due to field sampling techniques and those due to interpretation of the test from field response to added N.

The most commonly used procedures of NO_3^- determination among states using the test are the Nitrate Electrode and the Cadmium Reduction methods. These two methods are presented here. Any method used to determine NO_3^- should be scrutinized by using standard lab "check" soils, known additions and comparisons to reference procedures, such as steam distillation.

Phosphorus

All of the state soil testing laboratories in the North Central Region, except two, use the Bray and Kurtz P-1 procedure for phosphorus (P). The exceptions are North Dakota and South Dakota where soils are predominantly calcareous. Consequently, the Sodium Bicarbonate (Olsen) Method is used. Most of the states and provinces that border the North Central Region also use one of these two methods. Each state experiment station has developed correlations and calibrations for the soil conditions within its own state.

The Bray and Kurtz P-1 Test results are well-correlated with yield response on most acid and neutral soils in the region. This test is used for soils that contain small amounts (less than 2 percent) of dolomite or calcium carbonate. It should not be used for soils containing large amounts of lime. Since the phosphorus may be precipitated during extraction, the result is very low test values.

The Sodium Bicarbonate (Olsen) test for P is preferred for highly calcareous soils. The test results are well-correlated with crop response to P fertilization on both calcareous and noncalcareous soils. The Sodium Bicarbonate (Olsen) Test values are more highly correlated with yield response on calcareous soils than the Bray and Kurtz P-1 (1:10 ratio). In some cases, the correlation-to-yield was equal to or superior to the Bray and Kurtz P-1 on noncalcareous soils. If the Bray and Kurtz P-1 is the primary method used in the laboratory for extracting ortho-phosphate, the Sodium Bicarbonate (Olsen) Method should be used for highly calcareous soils that test very low in the Bray and Kurtz P-1 method.

In general, the Bray and Kurtz P-1 method will extract about the same amount of P as the Sodium Bicarbonate (Olsen) Method in the low range. The Bray and Kurtz P-1 method will extract more P in the medium range than the Olsen method. In the high range, except on highly calcareous soils, the Bray and Kurtz P-1 extracts more P than the Sodium Bicarbonate (Olsen) method. Each of these tests have a separate calibrations to yield response. Another extractant for determining P is the Mehlich 3 procedure.

This procedure exhibits a good correlation with the Sodium Bicarbonate (Olsen) method on calcareous and non-calcareous soils. However, Bray and Kurtz P-1 only correlates with Mehlich 3 on non-calcareous soils. Calibrations-to-yield response are more recent where they exist. Research is continuing to supply the calibration data. The response patterns of the Mehlich 3 extractant are similar to the Bray and Kurtz P-1 and the Sodium Bicarbonate (Olsen) tests. Mehlich 3 extracts more P from the soil than does the Bray and Kurtz P-1 or Sodium Bicarbonate (Olsen) tests on acid and neutral soils.

The Mehlich 3, Bray and Kurtz P-1 and Sodium Bicarbonate (Olsen) procedures can be used with various reducing agents. The procedures presented here use ascorbic acid with potassium antimony tartrate or an alternative using the Fiske-Subbarrow reducing agent (aminonaphtho-sulfonic acid). The Fiske-Subbarrow procedure is somewhat less sensitive and provides a wider range of soil test values without dilution. The ascorbic acid reducing method is welladapted to the Olsen method. A 1:10 soil-to-extractant ratio for Bray and Kurtz P-1 and Mehlich 3 (1:20 soil-toextractant ratio for Sodium Bicarbonate (Olsen)) should be maintained for either volume-to-volume or volume-to-weight ratios.

Bray and Kurtz P-1 Test

The Bray and Kurtz P-1 method of testing for "adsorbed" P was first published in 1945 by Roger H. Bray and Touby Kurtz of the Illinois Agricultural Experiment Station. Research showed the Bray and Kurtz P-1 to be best correlated with crop response to phosphate fertilizer. Most states implemented it into the routine soil test operation with various minor modifications. Surveys and sample exchanges conducted by the NCR-13 committee revealed diversity of phosphate concentrations extracted by the Bray and Kurtz P-1 procedure. Primarily, the differences were caused by extraction techniques that were not adequate.

Results of these studies are presented by Munter in his discussion on extraction techniques. A detailed study of the soil-to-solution ratio, type of extraction vessel, shaking speed and time, and the chemistry involved produced the following procedures. Using these procedures should provide the conditions necessary to obtain satisfactory replication of results among laboratories. The method detection limit is approximately 1.0 mg kg^{-1} (dry soil basis) and can be reproduced plus or minus 10 percent.

Equipment

1. No. 10 (2 mm opening) sieve
2. Standard NCR-13, 1 g and 2 g soil scoop
3. Automatic extractant dispenser, 25 mL capacity.(If preferred, pipettes are acceptable.)
4. 50 mL Erlenmeyer extraction flasks

5. Rotating or reciprocating shaker with a capability of 200 excursions per minute (epm)
6. Filter funnels, 9 to 11 cm
7. Whatman No. 42 or No. 2 (or equivalent) filter paper, 9 to 11 cm. (Acid resistant filter paper may be needed if using automated method of determining concentration by intensity of colour. Bits of filter paper may cause an obstruction in the injection valves.)
8. Funnel rack
9. Appropriate vials for colour development
10. Volumetric flasks and pipettes required for preparation of reagents and standard solutions; pipettes or a dilutor used for colour development
11. Photometric colorimeter (manual or automated) suitable for measurement in the 882 nm range (610 to 660 for Fiske-Subbarrow)
12. A computer or calculator, used for calculation of the concentrations of phosphorus in the soil

Extractant: 0.025 M HCl in 0.03 M NH_4F

1. Dissolve 11.11 g of reagent-grade ammonium fluoride (NH_4F) in about 9 L of distilled water.
2. Add 250 mL of 1.00 M HCl (previously standardized) and make to 10 L volume with distilled water.
3. Mix thoroughly.
4. The pH of the resulting solution should be 2.6 plus or minus .05. The adjustments to pH are made using HCl or ammonium hydroxide (NH_4OH).
5. Store in polyethylene.

Phosphorus Standards

1. Stock Standard Phosphorus Solution (50 ppm P)
   a. Dissolve 0.2197 g of oven-dried, reagentgrade potassium dihydrogen phosphate (KH_2PO_4) in about 25 mL of distilled water.

b. Dilute to a final volume of 1,000 mL with extracting solution. (If this solution is stored at 40°F, its shelf life should be approximately 6 months.)

2. Working Standard Solutions

a. Pipette appropriate volumes of 50 ppm stock standard P solution into proper volumetric flasks.

b. Use the extracting solution to bring each standard to the proper volume.

Potassium and Other Cations

Potassium (K), calcium (Ca), and magnesium (Mg) availabilities in soil are generally estimated by measurement of the water soluble and exchangeable forms. The amounts of K, Ca, and Mg in the soil solution are quite small relative to the amounts in the exchangeable form. Hence, the quantities of these three cations extracted in most soil test procedures are simply referred to as exchangeable K, Ca, and Mg. Available K levels in soils of the region are important for determining the appropriate rates of supplemental K to apply. Calcium in most North Central Region soils is rarely limiting as a plant nutrient.

The measurement of exchangeable Ca may be used with the measurements of the other exchangeable cations to calculate an estimate of the cation exchange capacity of soils and/or to calculate the percentage of base saturation as an index for the need to neutralize excess soil acidity. Magnesium deficiencies have occurred with sufficient frequency in the region to justify testing for Mg. A determination of available Mg will be helpful in deciding when to use dolomitic limestone.

The literature abounds with methodology used to measure exchangeable cations and cation exchange capacity. These methods were recently condensed. The reader is referred to these references for details of the most accurate and precise procedures for determining plant available and exchangeable cations. Soil testing or quick testing compromises some degree of accuracy for speed of determination. Therefore, standard or reference methods of soil testing have been developed to estimate nutrient availability. These estimates are then calibrated for recommendations based upon field trials with crop species of interest.

A reference method, such as the one described herein for potassium, must be calibrated for the soil/crop/environment continuum for which it is to be used. These calibrations carried out in the various states and/or soil association areas provide the data for interpretation of the soil tests in terms of fertilizer needs. Thus, if different crop/environment combinations give different yield responses in different soil association areas, different recommendations may result for the same soil test level. K extractable with neutral 1 M NH_4OAc has been calibrated with crop responses and supplemental K needs for the varied soils of the North Central Region.

The cation exchange capacity (CEC) of soils is important in determining the supplemental K needs and the appropriate quantities of soil-applied herbicides to use. The precise determination of CEC is time consuming. Soil testing labs in the region have determined that estimation of CEC by summation of exchangeable K, Ca and Mg and neutralizable acidity is acceptable for most soils. In Michigan, Warncke et al. found that CEC by summation is a good estimate of the actual CEC in acid, neutral and calcareous soils. Gelderman reports that CEC measures by summation may be inflated in calcareous soils by dissolution of $CaCO_3$ in the neutral 1 M NH_4OAc.

Sodium acetate is a better replacing solution to use in the determination of CEC in calcareous soils. The Mehlich 3 extractant has been adapted by many laboratories as a near-universal extractant. The results suggest that Mehlich 3 is a satisfactory extractant for K and for Ca and Mg on non-calcareous soils. These results show that either 1 M ammonium acetate or Mehlich 3 may be used to extract K. However, Mehlich 3 is not recommended as a substitute for 1 M ammonium acetate as an extractant for Ca and Mg from calcareous soils.

SULFATE-SULFUR

Testing the soil for plant available sulfur (S) and response to fertilization occurs mainly in the northwestern and western areas of the North Central Region. These areas have few industrial centers that produce gaseous or solid waste byproducts of S. Minnesota, Nebraska and Wisconsin report areas that show consistent plant response to S mainly on sandy soils low in organic matter. Soil test summaries often identify numerous soil tests of surface samples in the "low" category. However, field experiments show small or no crop response from S fertilization on the fields many of these samples represent.

Several states in the North Central Region and two Canadian provinces offer the plant available S soil test to the public. This interest in S tests reflects an increase in client demand. However, there is a need to compile information, develop techniques for analysis, and correlate the test results with response data. Three additional states are using the test only for research. Recommending applications of S on the basis that the amount applied is equivalent to crop removal or as "insurance" neglects contributions of S from the atmosphere, applied manure, the subsoil, and, for irrigated fields, any S in irrigation water.

Nature of Available Sulfur

Most of the S in the surface soils (95 to 99 percent in Iowa soils) occurs in organic combinations. Mineralization of organic S is an important source of plant available S. Plants can also absorb SO_2 directly from the atmosphere. The sulfate ion is the usual S form utilized by plants. Inorganic S occurs as the sulfate ion in combination with cations in soil solution (well drained, arable soil) or precipitated as a salt in conjunction with the existing cations (arid soil) and adsorbed by 1:1 clays and oxides of iron and aluminum. The adsorption increases as soil pH decreases below a value of 6.5.

The concentration of inorganic S determines the nutritional status of the crop since both the soluble and the adsorbed fractions are available. Soil test analysis usually measures the inorganic S or the inorganic and some organic S. The quantity of inorganic S present at any given time is small. It is continually undergoing changes due to: mineralization and immobilization by microorganisms leaching additions from the atmosphere in gaseous form or with precipitation and additions from the application of manure. Lower horizons of the soil profile and irrigation water may contain available S.

Measurements of inorganic S in surface soil at a given time do not always reflect all of the sources of readily or potentially available S. Other methods of predicting readily or potentially available S have serious shortcomings. Mineralization is slow and difficult to measure for prediction purposes. The atmospheric contribution fluctuates with the season and may be impractical to monitor except on a regional basis. Since the movement, retention and absorption of available S by plants occur predominately in the sulfate form, it is the fraction usually measured.

North Central Region S Tests

Sporadic responses to S, mainly by alfalfa, are recorded in parts of North and South Dakota, Nebraska, Minnesota and Wisconsin. Elsewhere in the North Central Region, S response has been minimal. Atmospheric precipitation, manure and subsoil are contributing S sources which may not be measured in the soil test. A 2-year average of 20 pounds S per acre per year was measured in precipitation in nonresponsive areas of Wisconsin and only half that amount in responsive areas.

A study of profile subsoil sulfate-S in six Wisconsin soils at eight sites gave amounts ranging from 11 kg/ha to 90 cm in a loamy sand to 179 kg/ha to 150 cm in a silt loam soil. The average profile S to 90 cm was 72 kg/ha. Sulfate-S through the profile correlated with organic matter in the 0 to 30 cm depth and was negatively correlated with pH. The sulfate-S concentration in medium and fine textured soils averaged from 3.3 to 7.4 kg/ha in the first foot, 5.9 to 13.1 kg/ha in the second foot and 8.9 to 18.5 kg/ha in the third foot increment based on data from 2,226 profiles submitted to the University of Wisconsin Soil and Plant Analysis Lab-Madison for residual NO3-N analysis in 1989-1991.

A previous manure application tended to slightly increase the average SO_4-S concentration in each foot increment whereas having grown a legume crop tended to result in less average SO4-S in each foot increment. These results help explain why response to S fertilizer is sometimes not obtained when analysis of surface samples would predict a response.

Considering S sources, a sample testing program similar to that for Nitrate-N (testing subsoil samples) and/or plant analysis may be the best predictors of the status of plant available S. Wisconsin is currently using a model to determine the need for additional S that includes S in precipitation, S released from soil organic matter, S from applied manure and S in subsoil, in addition to the sulfate-S soil test (16). The plant available S is the sum of these inputs and is expressed as the S availability index (SAI).

Micronutrients

Deficiencies of zinc (Zn), iron (Fe), manganese (Mn) and copper (Cu) are known to occur in the North Central Region. Of the total crop acreage in the region, however, only a small percentage is affected by micronutrient deficiencies. Zn deficiency has been recognized throughout the region; Fe, Mn, and Cu deficiencies are limited primarily to specific areas. Over the past three decades, micronutrient soil tests have improved markedly, in part because of improved instrumentation. However, many tests have been developed within a particular problem area and have not been extensively tested for their usefulness across wide areas.

More work is needed on relating micronutrient soil tests across geographic and climatic conditions. Most of the state soil testing laboratories in the North Central Region offer tests for one or more of the micronutrients. Cox and Kamprath found in their extensive review of many extraction procedures, which have been tried for the various micronutrients, that test results need to be supplemented with other information, such as pH, texture and presence of free lime for reliable interpretation.

Micronutrient contamination of samples can occur quite easily if care is not taken in collection and preparation of samples. Soil probes, sample containers and soil grinding equipment should all be checked for potential contamination before being used. The effects of sampling and sample storage time on test results have not been fully researched, so before incorporating the tests described in this section into a soil test program, one should study such effects for the specific soil/climatic conditions of a given region.

DTPA Extraction

The DTPA (diethylenetriaminepentaacetic acid) test, a nonequilibrium extraction developed by Lindsay and Norvell, has gained wide acceptance because of good correlation for Zn on calcareous soils and the potential for using the same extract for Fe, Mn and Cu. The DTPA test also shows considerable promise for use in monitoring cadmium, nickel and lead in soils that have received sludge applications. The DTPA test is presently being used as the soil test for Zn in Iowa, Kansas, Minnesota, Missouri, Nebraska, North Dakota and South Dakota. In addition, Kansas uses the DTPA test for Fe and Missouri uses the test for Cu, Fe, and Mn when requested.

Lindsay and Norvell showed that the amount of nutrient extracted by the DTPA method is affected by extractant pH, soil-to-solution ratio, chelating agent concentration, shaking time and extraction temperature. Subsequent work by others has shown that extraction intensity and sample preparation also affect the results. Control of extraction conditions is very important for comparable results among laboratories. Most laboratories are following the procedure as developed by Lindsay and Norvell, and deviations from this procedure must be carefully monitored to adjust the interpretation levels.

Equipment

1. Atomic absorption, inductively coupled atomic emission (ICP) or direct current plasma atomic emission spectrometers
2. Reciprocating or rotating shaker, capable of at least 180 excursions per minute (epm)
3. Burets or automatic pipettes, 50 mL Erlenmeyer flasks and filter funnels for extraction
4. Standard NCR-13, 10 g soil scoop, (0.85 cc/g) 5. Soil pulverizer with 10-mesh, stainless steel sieve checked for micronutrient contamination

Reagents

1. *Extracting Solution*: 0.005 M DTPA, 0.01 M CaCl2 and 0.1 M triethanolamine (TEA) adjusted to pH 7.3. For 18 L of solution, dissolve 35.4 g of DTPA in 268.6 g of TEA and about 200 mL of demineralized water and stir until dissolved. DTPA is slowly soluble in water but dissolves rapidly in the TEA-H_2O. Add 26.5 g of $CaCl_2.2H_2O$ to about 10 L of micronutrient-free, demineralized water; then add the DTPA/TEA mixture and bring to about 17 L with demineralized water. Adjust the pH to 7.3 using concentrated HCl and bring to the 18 L mark. Approximately 70 to 75 mL of HCl will be necessary.
2. *Stock Standards*: 1,000 ppm Zn, 1,000 ppm Fe, 1,000 ppm Mn, 1,000 ppm Cu.
3. *Working Standards*: Standards should be made up in the DTPA extracting solution. Concentration range for standards should cover Zn, 0 to 5 ppm; Fe, 0 to 10 ppm; Mn, 0 to 10 ppm; and Cu, 0 to 2 ppm.

Procedure

1. Air dry soil samples and crush to pass a 10-mesh stainless steel sieve.
2. Scoop 10 g of soil without pressing the soil against the side of the container. Firmly tap the handle of the scoop

three times with an 8 inch spatula and level off the soil by passing the spatula over the scoop, holding the spatula at a 90° angle.

3. Add the measured volume of soil to a 50 mL Erlenmeyer flask, tapping the scoop on the funnel or flask to remove all of the soil from the scoop.
4. Add 20 mL of extracting solution (1:2 soil-to solution ratio) to each flask and shake at 180 or more epm for 2 hours.
5. Filter through Whatman No. 42 filter paper or similar grade paper. Refilter if extract is cloudy. Those samples high in extractable Fe will have a yellow colour.
6. Carry a blank through the entire procedure with each run.
7. Read samples on the AA, ICP or DCP spectrometer unit using appropriate standards and instrument settings.
8. Report as ppm Zn, Fe, Mn or Cu in the soil.

Boron

Boron (B) is an essential nutrient to living plants. It has been characterized as a micronutrient because of the small quantity required to support optimum plant growth. B concentrations usually range from 5 to 80 μg per gram of dry plant tissue across plant species. The interval between deficiency and toxicity is narrow for most plant species. Some of the plants most sensitive to B deficiency are celery, cauliflower, cabbage, brussels sprouts, alfalfa, red clover, white clover, apple trees and pear trees.

Plants obtain B from soluble B forms present in the soil. According to Lindsay, H_3BO_3 is the predominant B species in soil solution. Only at pH above 9.2 is the H_2BO_3 species expected to become predominant in soils. Ideally, the test for soil B should measure the form of B that is most important to plants. A successful soil B test must, however, be able to measure the amount of B that is immediately available, as well as that potentially available to plant roots. The better the correlation between plant absorption of B and the measure of B in the soil, the more useful the test.

The B test must be sensitive enough to allow accurate measurements of concentrations (either high or low) which are important to the plant. In addition, the test must be free from major interferences caused by other chemical constituents in the soil extract. Berger and Truog divided soil B into three categories: total B, acid-soluble B (H_2SO_4), and watersoluble B. They concluded that water-soluble B correlated best with the incidence of black spot in garden beets. Work by Berger and Truog, as well as Starck, Truog and Attoe, showed that all the B added to a mineral soil could be recovered with a boiling hot water extraction.

'In 1966, Miljkovic, Matthews and Miller related the uptake of B by sunflowers from eight different soils to the concentration of soil B as determined by a hot water extraction. Next to watersoluble B, clay content had the most influence on B uptake. These two variables in a curvilinear regression accounted for 79 percent of the variability in uptake from cultivated surface soil samples.

Hot-water soluble B can be affected by many soil factors. Clays and oxides of iron and aluminum can fix B. Also, the soil organic matter content has been shown to be important, particularly for soils that are not highly cultivated. The absorption of hot-water soluble B by lucerne (alfalfa) was shown to be greater from coarse texture soils than from fine texture soils. A survey by Ouellette and Lachance revealed that when lucerne was the dominant plant species, B deficiency occurred more frequently on coarse texture soils than on fine texture soils. They concluded that about 0.8 lb of B per acre was necessary for normal growth of lucerne on fine texture soils compared to 0.5 lb of B per acre on coarse texture soils.

Variations in soil moisture and cultivation may also affect the amount of hot-water soluble B present. Work by Winsor showed that the concentration of hot-water soluble B increased as the soil moisture level increased. The increase occurred both in virgin and cultivated soils, but was much more in virgin soils. The soil texture in this research was fine sand. Methods that have commonly been used in the past to measure B have been those using quinalizarin and curcumin dyes.

Azomethine-H has been used to complex the B in plant tissue and soil extracts. Kowalenko and Luvkulich used a modified curcumin procedure and an acetate buffer extraction (pH 4.8) to measure available soil B. It must be emphasized that it is extremely important to use the instrument of detection that is recommended by the method. For example, if the method indicates that an inductively coupled plasma spectrograph (ICP) be used, then a colorimeter should not be substituted. Arbitrarily using another type of instrument can lead to serious errors in the analysis.

The Curcumin Method has generally replaced the Quinalizarin Method because concentrated sulfuric acid is not required for curcumin. Disadvantages of the Curcumin Method are that water must be evaporated from the sample, and a great deal of handling is thus required. An advantage of the Curcumin Method over the Azomethine-H Method is that of greater sensitivity. Methods that use ICP have greatly simplified the measurement of B.

Chlorides

Historically, soil chloride (Cl^-) analysis has been conducted primarily for the purpose of salinity characterization and irrigation management. However, recent research in the northwestern United States and in the northern Great Plains has indicated positive cereal responses to Cl^- additions. Studies in South Dakota have indicated that soil Cl^- level is a factor influencing the probability of obtaining a yield response to Cl^-. The procedures that will be discussed here are intended for determining Cl^- fertilizer needs rather than for salinity evaluation.

Therefore, detection of relatively low Cl^- concentrations is emphasized. Chloride is similar to nitrate in solubility and mobility in the soil. A 2 foot sampling depth was found to be superior to shallower or deeper depths for predicting wheat plant Cl^- concentrations in eastern South Dakota. Chloride is a ubiquitous ion, and precautions must be taken to avoid contamination during sampling and in the laboratory.

Many common laboratory reagents and cleansers contain Cl^-. Other possible sources of contamination include dust, perspiration, filter paper, glassware, water and paper bags. Plastic gloves should be worn when handling filter paper for Cl^- determination. Considerable flexibility exists in extraction techniques. Extractants that have been used include H_2O, 0.1 M $NaNO_3$, 0.5 M K_2SO_4 and 0.01 M $Ca(NO_3)_2$. Theoretically, these should give similar results; however, the method of determination used may make some extractants more convenient than others.

Time required for extraction appears to be similar to nitrate extraction. Gaines et al. showed that a 5 minute extraction on Georgia soils was adequate. Other investigators have adopted longer extraction periods of 15, 30, or 60 minutes. Minimum extraction times should be determined through recovery studies on the soils to be analysed. Comparison of soil-to-solution g to mL ratios of 1:2 or 1:2.5 showed much better precision levels than ratios of 1:4 or 1:5 using the Mercury Thiocyanate Method of Cl^- determination.

Several methods have been developed to determine determine Cl- in soil extracts. Many of these are not suited for routine soil testing and will not be discussed here. The procedures presented are those that have been successfully used in soil testing laboratories in the North Central Region. A comparison of the Mercury Thiocyanate and Potentiometric Known Addition Methods on a silt loam soil with a mean Cl^- concentration of 12 ppm found coeffecients of variance (CV's) of 9 and 24 percent, respectively. Precision values for most soil chloride methods are generally poor for samples with Cl- levels of less than 10 ppm. Typical CV's are 15 to 25 percent for such samples. Duplicate or triplicate analysis should be performed for these samples.

References

Gee, G.W., and M.E. Dodson., "Soil water content by microwave drying: A routine procedure," *Soil Sci. Soc. Amer. J.* 45:1234-1237, 1981.

Graham, E. R., "An explanation of theory and methods of soil testing," *Missouri Agric. Exp. Stn. Bull.*, 1959, 734.

Huffman, S.A. and K.A. Barbarick., "Soil nitrate analysis by cadmium reduction," *Comm. Soil Sci. Pl. Anal.*, 12(1):79-89, 1981.

Luebs, R.E., G. Stanford, and A.D. Scott., "Relation of available potassium to soil moisture," *Soil Sci. Soc. Amer. Proc.* 20:45-50, 1956.

Roth, G.W., D.B. Beegle, R.J. Fox, J.K. Toth, and W.P. Piekielek., "Development of a quicktest kit method to measure soil nitrate," *Comm. Soil Sci. Plant Anal.* 22:191-200, 1991.

Schofield, R. K., and A. W. Taylor., "The measurement of soil pH," *Soil Sci. Soc. Amer. Proc.* 19:164-167, 1955.

4

Managing Soil Architecture

Heat from the sun causes water at the surface of oceans, lakes and rivers to change into water vapour in a process called evaporation. Transpiration in plants is a similar process, in which water is absorbed from the soil by plant roots and transported up the stem to the leaves, from where it is released (transpired) as water vapour into the atmosphere. As the water vapour produced by evaporation and transpiration rises into the atmosphere, so the temperature decreases and water vapour changes into water droplets (condensation), which accumulate as clouds. Depending on their size, these may be released as rainfall. Once rainfall reaches the land surface it can infiltrate into the soil, run off over the surface as overland flow, or accumulate on plant leaves or in puddles from where it evaporates back to the atmosphere. A combination of these processes is commonly the case.

The rainfall that infiltrates into the soil forms part of the soil water, of which some may be used by plants for transpiration, some may return to the atmosphere through evaporation from the soil surface, and some—if sufficient infiltration occurs—may move beyond the rooting zone to the groundwater. The groundwater moves laterally and slowly towards the sea to complete the hydrological cycle, but part of it will seep into springs, streams, rivers and lakes on the way. In this way the groundwater maintains the water level in wells, and the continuity of river and streamflow during dry periods.

Rainwater that runs off the land moves rapidly downhill towards river courses, contributing to peak flows, and is of great concern. Runoff is not only a waste of rainfall that could have contributed to crop production and groundwater supplies, but it frequently causes floods or damage to roads and farmland, and erodes soil that is redeposited in river courses and reservoirs downstream. Groundwater is derived from rainwater that has infiltrated into the soil and drained beyond the rooting zone in excess of both the quantity needed for the crop or the vegetation and the water-storage capacity of the soil.

Groundwater moves very slowly through subsoil materials in the direction of the dominant drainage course. If its upper surface, the water table, does not sink below the level of the streambed, water is released to springs that feed streams and tributaries. This occurs throughout the year and in this way groundwater acts as a buffer in maintaining stream base flows and water levels in wells during dry periods. In soils with relatively impermeable subsoil layers beneath more highly permeable layers, perched water tables may develop above the groundwater, due to water being held up by the impermeable layers.

The water in a perched water table, sometimes referred to as interflow, will slowly move laterally and may emerge into stream courses or springs at lower elevations. It does not contribute directly to the groundwater. The presence of groundwater or a perched water table is indicated by saturated soils, and usually by a dominance of light grey, bluish-grey, bluish or greenish colours. These colours are typical of certain iron compounds that only form in waterlogged soils where oxygen is lacking.

The amount of rainfall that percolates beyond the lower limit of the rooting zone towards the groundwater will depend on the amount of water used for transpiration by the crops or vegetation. For a particular climate and soil type, forest transpires more water than grassland, which generally uses more water than crops. The high water use by forest is due to its generally greater transpiration rate, the longer period of transpiration compared with crops, and the deeper roots enabling it to absorb water from greater depths.

Changes in land use can therefore affect the quantity of water transpired and hence the quantity reaching the groundwater. Replacing forest vegetation with grassland or annual crops may increase deep drainage and so provide higher base flows in streams and rivers. Changes in soil management can also affect the quantity of deep drainage replenishing groundwater. The introduction of poor management practices that increase the proportion of rainfall lost as runoff will reduce base flows and increase peak flows and the incidence of flooding.

Conversely, an improvement in soil and nutrient management will lead to higher grain and foliage production, higher transpiration rates, and hence less recharge. In order to consider rainwater for plants and for groundwater as parts of a sequence, it is important to have a mental picture of its journey. After passing through the atmosphere in response to gravity, water from rain or irrigation travels to some or all of the following destinations.

Legend:

- — Direct evaporation from wetted leaf surfaces.
- — Surface runoff/stormflow.
- — Direct evaporation from the soil surface.
- — Plant-available soil moisture within root-range of existing weeds, crops, trees.
- — Soil moisture within root-range of existing plants but held at tensions unavailable to them.
- — Soil moistures held at all tensions, but below root-depth of existing plants.
- — Water not captured by roots and small pores, moving to groundwater and streamflow.
- — Leakage to deep groundwater beneath catchment floor.

Management of the soil can significantly affect runoff; direct evaporation from the soil surface; the amount of soil moisture available to plants within range of their roots; and the depth to which roots can penetrate. How much water reaches each of these destinations over a given period depends on the physical condition of the soil and its influence on infiltration and runoff, and on the atmospheric conditions as they affect evaporation and transpiration.

Catchments and Watersheds

Water caught by a catchment will flow towards the lowest point at the outlet, where it may join water emerging from other catchments. The outer boundaries of a catchment are defined by ridgelines along the crests of the surrounding uplands. From the sides of a valley surface runoff tends to flow perpendicularly to the slope from crest to streamline.

A watershed is the area of land dividing two streamlines. Water moves away from the crest line towards the streamlines on either side. Thus a hill slope can be considered as either the inner slope of a catchment or the outer slope of the watershed. Catchments and watersheds are indicated on maps by the contour lines and by the course of drainage lines. Underlying geological formations, together with weathering and uplift processes, affect the form of landscapes. They influence the steepness or shallowness of slopes, whether the streamlines are of relatively sinuous shape or with abrupt changes of direction. The flow of water along the streamlines tends to cut the heads of streamlines back into the underlying materials.

For the purposes of enabling rainwater to soak into the soil and controlling the rate of flow of any excess runoff, we can subdivide a given catchment into a more detailed hierarchy of catchments, in which the smallest subdivisions may be measured in square centimetres, the larger in hectares, within catchments of square kilometres. Rainfall entry into the soil depends on the porosity of the soil at any scale, while management of runoff and erosion across the surface also depends on any physical works that may be constructed when rainfall rates exceed even the best infiltration rate.

Overlapping pairs of vertical aerial photographs viewed with a stereoscope provide a three-dimensional view of the landscape and surface features. Plate 11 is a stereogram that shows a layout of roads along topographic crests and of conservation banks close to contours that has been designed in conformity with the natural catchments of the landscape. This pattern provides a framework within which planting rows will have been contour-aligned relative to the conservation banks.

The maintenance of soil porosity by mulch cover will allow the highest proportion of rainfall to be available as soil moisture for the crops and groundwater for the streams. A stream catchment may be large or small, of steep or shallow slope and composed of natural subcatchments and then of field catchments. Plate 12 shows two of these field catchments, which also form the left-hand side of a watershed whose crest runs along the ridge seen at top right.

In cropping agriculture the next smaller subdivision is the bund catchment, between any pair of physical conservation banks with its ridges and furrows (formal or informal) along planting rows. The physical conservation banks, vegetated with a fodder grass, subdivide the field catchment as well as separating the bund catchments. The aim of these structures is to conserve water and soil, though their effects on yield are disappointing. Their most significant function is to provide guidelines for contour planting of crop rows.

The smallest subdivision to trap rainwater and give it time to soak in is the microcatchment with its mulch. In forestry, because the young trees are more widely spaced, the same effect can be obtained by a set of half moon shaped microcatchments, one at each planting position. Within this overall framework, the key to infiltration is to keep the soil porous with a cover of crop residues, which prevents damaging raindrop impact and provides a substrate for soil organisms.

The conservation effects of forests are due not so much to the presence of the trees themselves but to the litter of fallen leaves, twigs and branches, plus any low-growing vegetation. If the soil surface has not been damaged by trampling, less rainwater will run off and more will infiltrate into the soil.

How much plant-available soil moisture remains at a given time depends on the texture and porosity of the soil, the previous volume of soil moisture, the volume removed by direct evaporation, evapotranspiration and deep drainage. Irrigation (if available) is normally required when about two thirds of the available water—between field capacity (FC) and permanent wilting point (PWP)—has been depleted.

If irrigation is not an option, it makes sense to manage the soil to develop and retain a maximum amount of soil pores of a wide range of sizes. This will maximise the capacity for water retention and enable plants to withstand drought for longer periods. Loam textures generally have the largest available water capacity, while sand on the one extreme has a small available water capacity, as does clay at the other. Available water capacity coupled with soil depth determines the volume of water usable by plants at a particular site. This is illustrated by comparing relevant characteristics and consequent amounts of available water for two soils on which tea is grown, one at Timbilil in Kenya, the other at Marikitanda in Tanzania.

Flood flows in streams and rivers, which rise quickly after heavy rainfall derive mostly from rapid overland flow of water. Flood flows are often muddy with eroded materials. Clear streamflow originates from rainwater, which has infiltrated the soil and percolated through pores of a range of sizes at different slower speeds. The Plates 19 - 22 show streamflow during the rains from a cultivated catchment without any effective conservation measures and a nearby forested catchment, both on the slopes of the same mountain, within 1 km of each other.

The clear water, which runs throughout the year, more voluminous in the rainy than in the dry season, has percolated through the litter of fallen leaves and branches on the forest floor, which both protects the surface from rainfall impact and nourishes the soil organisms that maintain soil porosity. This water has travelled the slow route down through the soil to the groundwater, which moves into the stream via springs and seepages along the streambanks.

As dug wells provide direct access to shallow groundwater, on which many rural communities rely, it is important that enough rainwater penetrates and pass through the soil to replenish groundwater. Exceptional rainfall such as during typhoons or hurricanes on already saturated soil can result in floods and erosion that change the landscape, no matter how well the land and the crops, grassland or forest are managed.

Pore Spaces in Soils

Although we generally think of soil in terms of its solid parts, i.e. the sand, silt, clay and organic matter, it is the spaces between these solid particles that are as important as the solid particles. This is because it is the spaces where all the action takes place, just as in a house, where all the important activities occur in the rooms rather than in the walls and floors. It is therefore the architecture of the soil that is important. The pore spaces in a soil vary in abundance according to the type of soil and how it has been managed.

Soils under natural vegetation generally exhibit high porosity because of high biological activity and lack of interference by man. Consequently they have superior physical qualities compared with most soils used for crops or grazing. Plate 25 illustrates the contrasting porosity in forest and cultivated soils. Pore spaces in soils vary in size, and both the size and continuity of pores have an important influence on the types of activities that occur in soil pores.

Pore sizes from 0.0002 to 0.05 mm diameter retain water that can be absorbed by crops and are referred to as storage pores, whereas smaller pores hold water too tightly for plants to be able to extract it. Pores larger than about 0.05 mm diameter, referred to as transmission pores, allow water to drain through the soil and enable air to enter the pores as the water drains out. Pore spaces are also needed for roots to freely penetrate soils in order to take up nutrients and water.

The sizes of roots vary with the type of crop, but the smallest roots, apart from root hairs, have diameters of 0.1 to 0.3 mm and so soils must have pore spaces of at least this size if the smaller roots are to penetrate freely. In most soils roots grow partly through existing pores, the transmission pores, and partly by moving aside soil particles. Roots can only force their way into smaller pores if the soils are sufficiently compressible; the compressibility of soils increases with increasing water content, since water provides a form of lubrication between soil particles.

Amount of Water in Soil

The amount of water present in a soil, which is available for crop production, will depend on how much of the rainwater remains in the soil after the losses by runoff, evaporation, and deep drainage. The amount of rainfall that reaches the groundwater and thus contributes to water security, will depend on the extent to which the rainfall infiltrating the soil is in excess of that needed to replenish the soil's water holding capacity and satisfy the transpiration needs of the crops.

Good rainwater management aims to maximize the amount of rainwater that enters the soil, and to make best use of it while it is there for use by crops and for recharging the groundwater. Any truly unavoidable surface runoff is conducted away safely in such a manner that it does not cause erosion problems. When a well drained soil is saturated to the limit of its rooting zone, the rainwater that does not drain out of the root zone within 48 hours will be retained in soil pores smaller than about 0.05 mm diameter.

The quantity of water retained after 48 hours corresponds to the soil's field capacity (FC). The forces (or suctions) with which this water is held will vary according to pore size. The largest pores still to retain water will hold the water at about a tenth to a third of the pressure of the atmosphere, depending on what suction corresponds to the soil's FC; this will vary with soil type and depth of the water table.

The maximum suction most crops can exert to withdraw water from soil varies with the crop, but the generally accepted value is equivalent to about 15 times the pressure of the atmosphere. This is approximately equivalent to the pressure that would be experienced when supporting a tonne weight on the palm of the hand.

When soil water has been exhausted down to 15 bars, the water remaining in the soil will be that stored in pores smaller than 0.0002 mm diameter, and will correspond to the soil's PWP. Water held at suctions greater than the PWP is not available for plant growth. Consequently, it is water held between FC and PWP which can be used by crops for transpiration, and is termed the soil's Available Water Capacity (AWC). However, after a heavy rainstorm some of the water in excess of the soil's FC may be used by a crop while this excess water is percolating through the rooting zone.

The available water held within the range FC to PWP is retained with different strength, and about a third of it is not easily or rapidly available to crops, especially if the crops are transpiring strongly. The higher the transpiration demand, the more available the soil water must be to avoid crop water stress. In contrast, for a slowly transpiring crop even water held at higher suctions can be used without causing stress. The maximum amount of available water that a soil can retain will vary with the soil's texture, organic matter content, rooting depth and structure.

Soil organic matter is particularly important in that it can retain about 20 times its weight of water. Organic soils and medium textured loamy soils with high contents of very fine sand and silt generally have the highest AWCs, clayey soils intermediate values, and soils with high contents of coarse sand the lowest AWCs. The stone content of soils can also be very important depending on the nature and abundance of the stones. Some ironstone gravel > 2 mm diameter can contain more than 20 percent water (m^3/m^3) at FC and porous limestone and chalk can also make significant contributions to the AWC of a soil.

In contrast, a high content of non-porous stones will greatly diminish the AWC of a soil. For any given soil, the greater the rooting depth, the larger will be the quantity of soil water available to the crop. This is particularly important for annual crops as they have less time to develop deep and extensive rooting systems than perennial crops. The available water capacity may influence the length of growing period for crops grown on that soil. Soils of high available water capacity will permit longer growing periods because of their ability to provide greater quantities of stored water during dry periods than soils of low available water capacity. Shallow soils have little available water, and even in wet years they will be unable to benefit by storing any more water.

Infiltration of Rainwater

In most areas where water shortages occur, maximising the infiltration of rainfall into soil is indispensable to achieving food and water security. Land management should encourage infiltration as opposed to runoff. Exceptions are where rainwater harvesting is necessary for crop production and where high infiltration can lead to risks of landslides or other forms of mass movement. The amount of rainfall that infiltrates will be governed by the intensity of the rainstorm in relation to the soil's infiltration rate.

Excessive tillage and loss of soil organic mater often result in reduced infiltration rate due to loss of surface porosity. When storm intensity is greater than soil infiltration rate, runoff will occur, resulting in a waste of water that should have been used for crop production and for recharging the groundwater. The rate at which rainfall infiltrates into soil is influenced by the abundance, stability and size of the pores at the soil surface, their water content and by the continuity of the transmission pores into the rooting zone.

In many soils the number of surface pores is rapidly reduced by the impact of raindrops, which break surface soil aggregates into small particles that clog surface pores and form surface seals or crusts with very few pores. The destructive raindrop action is avoided where there is a protective cover of crop foliage, residues, mulches or even weeds at or over the soil surface.

Other factors that can reduce the number, proportion and continuity of transmission pores are traffic by machinery, humans and animals, which destroys large pores by compaction, and tillage which disrupts the continuity of transmission pores through the smearing and compression of pores during plough pan formation in the subsoil. Infiltration rates are also affected by (1) the quantity of water present in the soil at the time of the rainstorm, which will depend on when the last rainstorm occurred and the permeability of the soil, and (2) the soil's capacity to retain water, which will vary with soil depth, stoniness, and texture.

Percolation of Rainwater

When a heavy rainstorm falls on a well-structured soil, rainwater percolates down through the dry soil as a wetting front, temporarily saturating the soil and displacing air. This is accompanied by the rapid drainage of water from the larger pores through gravity and the pressure of the mass of rainwater above. These larger pores exert only small forces of attraction on soil water. After about two days of drainage field capacity will have been attained and air will have re-entered the larger pores. In poorly structured soils, rainwater will drain much more slowly. Drainage often continues for several weeks depending on the depth to the slowest horizon and the continuity of the larger pores with depth. In fine textured soils with cracks drainage water will flow down through the cracks in heavy rainstorms before the soil is saturated and while parts of the soil profile may still be dry.

If the drainage water subsequently enters a smaller pore while passing through the soil, it will be retained, otherwise it will continue until it reaches the water table and contributes to the recharge of groundwater. Once the drainage water has been lost from the rooting zone, further water movement within the root zone is slow and is referred to as capillary movement. This movement is caused by forces of attraction, known as surface tension forces, which are exerted by soil particles on water. This movement can occur in any direction and includes the upward movement of water from water tables. Surface tension forces pull water into pores within the soil and the smaller the pores the more strongly the water is attracted and held.

Loss of Water Vapour from Soil Surfaces

Water is also able to move through soils as water vapour. The most important example of this is the loss of water vapour by evaporation from soil surfaces. This occurs when the concentration of water vapour in the soil close to the surface is higher than that in the atmosphere immediately above. Water vapour will then move from the soil into the atmosphere. The drier and hotter the atmosphere compared with the surface soil, the greater will be the rate of evaporation from the soil, provided sufficient water can be supplied to the surface by capillary movement from below. Fine textured soils have an abundance of small pores and so more capillary movement of water to the surface will generally occur in fine textured than in coarse textured soils.

Water Movements in Plant

Crops use large quantities of water, which under rainfed conditions come entirely from water in the soil, which in turn is derived from rainfall that has infiltrated the soil. A maize crop may use 400 to 750 mm of water depending on the rainfall and evaporation conditions. This corresponds to 4 000 to 7 500 cubic metres of water per hectare over the growing season. Almost all the water absorbed from the soil by crop roots passes up through the stem into the leaves, where it evaporates and passes into the atmosphere in a process known as transpiration. This process accounts for almost all of the water absorbed by plant roots.

Transpiration is essentially the same as the process of evaporation. Evaporation is what happens when a bowl of water is left in the sun. The liquid water disappears as it becomes converted into water vapour, and the higher the temperature, the drier the air, and the greater the wind speed, the greater will be the rate of evaporation. Evaporation occurs whenever water is exposed to the atmosphere, i.e. from lakes, rivers and puddles, and from the raindrops that accumulate on a leaf after a rainstorm.

To ensure an efficient uptake of sufficient water by crops it is important that the crop roots are well distributed and able to penetrate deeply into the soil. As a soil dries out from the surface downwards, so roots in the deeper layers tend to increase in number to compensate. When soil water reaches the surface of a root or root hair, it moves across the root into the xylem, which contains narrow tubes running through the root and extending up through the stems into the leaves.

On reaching the leaves, water passes from the xylem into leaf cells where it evaporates into air spaces within the leaf. These air spaces are saturated with water vapour, and are connected to the normally drier outside air by very small openings in the leaves called stomata. During the day the stomata open, which allows carbon dioxide to enter the leaf. Sunlight is used to make sugars within the plant: a process known as photosynthesis. Part of the sugars are used to produce energy by a process called respiration, part are converted into substances forming the various plant organs.

Photosynthesis occurs only during daylight, whereas respiration occurs all the time. When the stomata open to allow carbon dioxide to enter, water vapour escapes into the drier air outside. For transpiration to occur there must be a continuous supply and movement of water from the soil to the plant to the atmosphere. The driving force responsible for this movement is the same as for evaporation, and can be simply described as the tendency for water to move, either as a liquid or a vapour, from where it is more abundant to where it is less abundant.

In transpiration, water vapour moves from the very humid air spaces within the leaf into the drier atmosphere outside the leaf where the water vapour concentration is lower. The movement of water vapour out of the leaf creates a suction on the water in the leaf cells, the xylem, the roots and the soil, so water moves into the root, up the xylem and into the leaves, to replace that which has been lost from the leaves. In addition to the transpirational suction, which causes water to move from the soil into the root, there is another force attracting water into the root known as osmosis.

In osmosis, water moves from where it is more pure to where it is less pure across a semi-permeable membrane. A semi-permeable membrane is a very thin skin, which has pores large enough for water to pass through into the root but not large enough for dissolved salts to pass out of the root. Water therefore passes from the soil where the water is more pure across the root surface into the root where the water is less pure.

Water Stress

Many areas with low and erratic rainfall where crop water stress is common are also deficient in nutrients, and the lack of nutrients is frequently the second most limiting soil factor. An interaction often occurs between soil water and nutrients, which means that soil water can influence the availability of nutrients, and the availability of nutrients can influence the uptake of soil water and a crop's resistance to drought. Thus, both factors can influence each other. Plants contain a certain amount of water within them, which acts as a buffer against times of water shortage, but the amount is too small to last long.

In contrast, plants store sufficient quantities of nutrients within their tissues to provide a buffer for longer periods when nutrients are not being absorbed. Consequently, water deficiencies become more quickly apparent and damaging than nutrient shortages. This suggests that conserving water may often be of prior and quicker benefit than attempting to conserve soil particles per se. In addition, a lack of water also reduces the uptake of nutrients by a crop. This is largely because nutrients can only move to roots through water films within the soil, and so there must be continuous water films connecting the nutrients with the roots.

A lack of soil water continuity, due to drought for example, will severely reduce the rate of nutrient uptake by crops. A lack of soil water will also diminish nutrient availability by reducing microbial activity, which is responsible for the liberation of nitrogen, phosphorus and sulphur from soil organic matter. When there is a drought it is the surface soil, which dries out first, and so while a crop may still be able to absorb water from the subsoil, it may suffer from a lack of nutrients. A lack of available nutrients in the soil can restrict crop water uptake, especially when nutrients are limiting root development. This occurs most often in soils that are deficient in phosphorus. Applying P fertiliser to P-deficient soils will often promote root development, and as a result crop water uptake. Consequently the beneficial effects of applying P fertilisers are often relatively greater in seasons of lower rainfall than in those of higher rainfall.

The effects of drought and nutrient availability on crop yields are difficult to predict, because the effect will depend on when the water or nutrient shortage occurs in relation to the crop's stage of growth, and its needs and sensitivity to a lack of water or nutrients at that time. It is therefore often difficult to assess which factor, e.g. water or nutrients, is the more limiting to yield.

The most limiting factor can vary from season to season depending, for example, on when water shortages occur, and even during a season there will probably be periods when water is the main limiting factor, and other periods when nutrients are most limiting. Water shortages often affect whether or not there is a response to fertilisers, and how much fertiliser should be applied. This is particularly common with N fertiliser, where the optimum response is frequently higher in good seasons than in poor seasons. For example, when there is no shortage of soil water an application of 40 kg/ha may prove to be the optimum application rate, but when water is lacking only 20 kg/ha may be the optimum amount to apply.

This creates difficulties in rainfed agriculture: since it is not possible to reliably predict the distribution and amount of rainfall, farmers cannot know how much fertiliser to apply. One approach that can help to overcome this problem is to apply a modest amount of N fertiliser at the beginning of the season assuming low rainfall, and then to apply additional quantities N later if the season appears promising.

Restricted Rooting

The most common cause of restricted rooting is physical restriction due to soil compaction, which results in the collapse or diminution of pore spaces and a localised increase in bulk density. Once pores have been compacted to less than about 0.2-0.3 mm diameter, it is difficult for crop roots to freely penetrate the soil. Although the strength of compacted layers decreases as soil water content increases, a high water content can quickly limit the supply of oxygen to roots, so that roots then become restricted by a lack of oxygen. Certain crops, such as cotton and sunflower, appear to be more susceptible to restricted rooting from compacted layers than others.

Compaction often reduces pore sizes sufficiently to inhibit root penetration but not sufficiently to affect the drainage of water through the soil. Pores of 0.2-0.3 mm diameter can restrict roots but water can drain under gravity through pores as small as 0.01 mm diameter. In mechanised cropping systems the continual use of tillage implements, especially disc ploughs, disc harrows, mould-board ploughs and rotovators, over long periods of time frequently results in the formation of dense plough pans containing few pores large enough to be penetrated by crop roots.

The plough pans develop just below the depth to which the soil is tilled and often have smooth upper surfaces with sealed pores, caused by the smearing action of mould-board ploughs. The degree of compaction depends on the pressure exerted by the implements on the soil. Land preparation when soils are wetter than the optimum moisture content for tillage promotes soil compaction, because the soils are then much more compressible. This is particularly likely to occur on soils that have deficient drainage, or are difficult to till in a dry state without pulverising because of their very hard consistence. Compaction is also more likely when farmers use many passes to prepare the seed bed, or when they have only limited tractor power available and are unable to use wide sets of equipment and therefore produce compacted wheel ruts at closer spacing across the field surface.

Compaction can also develop in the subsoil from the passage of heavy machinery such as combine harvesters and lorries loaded with grain, especially in wet conditions. The degree of compaction will depend on the total axle load of the machinery. Soil compaction can also develop from hand tillage. Thin hoe pans just 2-3 cm thick can develop just below the depth of hoe penetration and thus restrict root penetration. When mounds or ridges are formed every year, the combination of hoeing at the same depth and the traffic of people within the furrows during wet conditions may accentuate the compaction.

A similar effect to that of compaction can occur when structurally unstable soils, known as hardsetting soils, slump on becoming saturated by intense rainstorms to form dense layers. On drying the dense soil layer becomes very hard and restricts root penetration. Restricted rooting may also be caused by naturally occurring dense horizons containing few pores large enough for roots to penetrate. These horizons may be found in soils formed from river, lake or volcanic sediments and in semiarid and arid areas where chemically cemented calcrete and gypsic horizons are formed.

Indicators of Restricted Rooting

The most obvious indicator of restricted rooting when a crop is present is the distribution of the crop roots. When roots are physically restricted by a dense layer containing few pores suitable for root penetration, individual roots often develop characteristic growth patterns immediately above the restricting layer. The most common of these is the abrupt change in the direction of growth from vertical to horizontal, and a thickening of roots that do manage to penetrate the restricting layer just above the upper boundary of that layer.

In mechanised agriculture, plough pans are usually formed at 12-30 cm depth, depending on the implement used and its normal working depth. Naturally occurring dense layers may occur at any depth. The optimum time to observe roots is after flowering when most of the roots will have largely completed their growth. When no crop is present, it is much more difficult to identify the existence of potentially root-restricting layers in a soil. However, the rooting pattern of mature weeds, either rooted or uprooted, that remain in the field after the crop has been harvested can be used to reveal the existence of a root-restricting layer.

When neither a crop nor weeds are present, the presence of a dense soil layer of high strength and containing very few visible pores will often be a useful indicator. The presence of dense layers is often revealed when digging by the abrupt increase in resistance to the spade or hoe when the restricting layer is reached. However, sudden increases in soil resistance can also be experienced when the soil changes from moist to dry. To avoid this problem, it is advisable to wet the soil to 30 cm depth two days prior to carrying out the field examination. Physically restricting layers can be identified by the scarcity of visible pores. The smallest pore visible to the naked eye (0.1 mm diameter) coincides reasonably well with the smallest pores into which the seminal roots of cereals (0.1 to 1 mm) and the tap-roots of dicotyledons (0.3 to 10 mm) can penetrate.

When the density of visible pores observed in fragments of the dense layer from a soil pit is less than about six in an area of 10 cm × 10 cm, root restriction is likely to be severe, and responses to breaking up the restricting layer are likely. Other indicators of potentially root-restricting layers that can be used in the field in the absence of a crop are soil strength determined with a penetrometer, and soil bulk density determined from undisturbed soil samples of a known volume. Critical penetrometer resistance and bulk density values at which the roots of most annual crops are restricted have been established for soils of different textures.

References

Brown R. B., "Soil Texture," *Fact Sheet SL-29*, University of Florida, Institute of Food and Agricultural Sciences, Retrieved on 2007-12-02, September 2003.

Buol, S. W.; Hole, F. D. and McCracken, R. J., *Soil Genesis and Classification*, First, Ames, IA: Iowa State University Press, 1973.

Voroney, R. P., "The Soil Habitat in Soil Microbiology," *Ecology and Biochemistry*, Eldor A. Paul (ed.), 2006.

5

Soil System Management in Temperate Regions

All generalisations about soil systems have significant exceptions, but two generalisations that differentiate temperate from tropical soils are quite tenable. First, because temperate soils are seasonally cold, during a significant portion of the year, plant growth and soil biological activity are low or nil due to suboptimal or freezing soil temperatures. Seasonality with its temperature fluctuations results in important changes in the chemical and physical soil environment.

Freeze-thaw cycles accelerate rock weathering and the breakdown of soil aggregates, for example, and chemical as well as biological reactions that affect mineral weathering, chemical solubility, and other soil chemical properties occur more slowly in winter months. As discussed below, this seasonality provides both challenges and opportunities for effectively managing soil fertility. A second generalisation concerns soil mineralogy and its impact on soil chemistry. Agricultural soils in temperate regions are more likely to be geologically young in comparison to large regions of the tropics.

Some tropical soils are also young, especially those developed from geologically recent volcanic and alluvial deposits, but most are not, and this has important implications for soil fertility. In young soils, such as those recently glaciated or formed from windblown loess, primary minerals have weathered little, and the electrical charge system is largely permanent, with base cations such as Ca^{+2}, K^{+}, and Mg^{+2} common.

In older soils, weathering will have removed most of the 2:1 layer-silicate clays, and electrical charges result mainly from the protonation and deprotonation of surface hydroxyl groups. The ion exchange capacity of older tropical soils therefore depends very much on soil pH. When pH is low, cation exchange will be negligible, and many ions important for plant growth will be in low supply and easily washed from the soil by percolating rainfall.

In contrast, the charge system of younger soils is more durable because it mostly results from the crystal lattice structure of 2:1 layer-silicate clays and is thus more impervious to changes in soil pH and soil solution composition. Although all soils contain both permanent and variable charge surfaces, most are dominated by one or the other charge system, and this has a significant impact on nutrient mobility and availability. In general, the permanent charge system that dominates most temperate region soils provides these soils with greater chemical and structural resistance to the deleterious effects of chronic disturbance that is typical of mechanised agriculture.

One sometimes reads of other temperate vs. tropical soil differences such as regional differences in soil organic matter (SOM) or soil biodiversity. These are not differences that are regionally inherent. Many tropical soils can have native SOM stores equal to those in temperate regions, so such generalisations do not hold up. Likewise, it is difficult to generalise about soil biodiversity because we know so little about it.

We know, for example, that 1 g of soil can contain $>10^9$ microbes representing >4000 different, mostly unidentified species, while a liter of soil can contain hundreds of different species of soil fauna. On average, about 20% of the organic matter in arable soils is living biomass, yet very little of this can at present be identified by species. In neither temperate nor tropical soils do we know >1% of the soil biota, so generalisations are hard to substantiate. Moreover, the relationship of this biodiversity to ecosystem functioning is in any case not yet documented. So, scientists and practitioners are both operating with little certain knowledge about the specific organisms present in soil.

Challenges to Soil Fertility

The inherent fertility of many temperate-region soils is high. In comparison to highly weathered tropical soils, many soils in temperate regions can withstand years of crop production following their conversion from natural vegetation. Eventually, however, soil nutrient stocks decline and soil structure degrades, and most temperate-region cropping systems now owe much of their present productivity to external subsidies, which enhance or compensate for lost ecological services.

To bring soil to its full fertility and to sustain this depends on the satisfactory resolution of two major challenges: the restoration and maintenance of SOM, including its all-important living fraction; and the development of nutrient-efficient, and especially nitrogen-efficient, cropping systems. Other challenges are also important—erosion control; water conservation; nutrient losses to groundwater, surface waters and the atmosphere; and pathogen suppression, among others—but in most landscapes they remain secondary.

Soil Organic Matter Restoration

The loss of SOM, sometimes referred to simply as soil carbon loss, is common to almost all field-crop production systems. The principal cause of SOM loss is accelerated microbial activity as agronomic activities in general, and cultivation in particular, stimulates microbial respiration of soil organic carbon, the foundation of organic matter. This rapid turnover of SOM is the foundation of soil fertility in low-input cropping systems, since as microbes consume carbon they release nitrogen and other nutrients to the soil solution where nutrients become available to plants.

Typically 40-60% of a soil's organic carbon stores are lost in the 40-60 years following the initial cultivation of a temperate region soil. This occurs even more quickly with cultivation in the tropics. Restoring lost SOM and tempering its turnover is thus a major goal of biologically-based agriculture. The reasons for accelerated microbial activity are complex and related to a number of factors. Chief among them is the breakdown of soil aggregates, small particles of soil (0.05-8 mm) that protect carbon molecules from rapid microbial consumption.

Carbon particles inside aggregates exist in an environment very different from the bulk soil environment: certain soil organisms may not be present, and the activity of those that are present is likely be restricted by low oxygen availability. Oxygen diffuses very slowly into aggregates, with the result that the oxygen consumed by microorganisms is not quickly replaced, and the interiors of aggregates thus tend to be anaerobic to a much greater degree than bulk soil. Cultivation breaks apart aggregates, especially the larger ones, exposing trapped organic carbon to aerobic microbes that easily respire it to CO_2.

Much of the increase in atmospheric CO_2 starting early in the 19th century was the result of pioneer cultivation. This stimulated microbial activity and the turnover of active organic matter pools formerly in aggregates. The basis for soil carbon sequestration as a CO_2 mitigation strategy is recovery of this lost soil carbon. While this recovery will contribute to greenhouse gas abatement, it will also improve soil productivity and increase microbial biomass and nutrient availability.

Aggregation remains low following cultivation because the microbial processes that stabilise soils are disrupted at the same time that aggregates are more exposed to physically destabilising processes. Microbial production of polysaccharides, humic substances, and aliphatic compounds that promote particle binding and aggregate stabilisation invariably decrease following the start of cultivation. Additionally, extensive networks of fungal hyphae that enmesh soil particles and provide a framework for aggregate stabilisation are shattered by tillage. These hyphae are also sensitive to changes in residue placement following agricultural conversion.

Increases in the physical forces that destabilise aggregates following cultivation are mostly related to changes in soil water dynamics. Cultivated soils are bare much of the year. During these periods, raindrop impacts result in greater disruption of aggregates at the soil surface, and without transpiration these soils will be wetter for more of the year. Generally, with increasing water content, aggregate structure decreases and dispersed clay increases. Bare soils are also more exposed to freezing, which has particularly damaging effects on soil structure since as soil water freezes and expands, it moves into pores and fracture planes between particles, driving them apart.

Structural breakdown of cultivated soils leads quickly to wind and water erosion, to substantial and permanent losses of soil carbon, and ultimately to reduced productivity. Usually not all carbon is lost from soil even after decades or centuries of plowing. However, what remains is carbon that is relatively unavailable to microbes because it is chemically resistant to microbial decomposition or tightly bound to clay particles—what soil biologists call slow or passive carbon—plus whatever carbon has been recently added as crop residue. These fractions provide a very different soil habitat than before, bereft of many of the benefits of abundant SOM and biota. There is less water-holding capacity, less porosity and aeration, lower infiltration, and a diminished buffer of biologically available nutrients.

Moreover, organic matter itself—even in permanent-charge soils—provides significant cation exchange capacity, which helps to hold biologically important cations against leaching loss. Soils impoverished in carbon will thus be impoverished in biological activity and in the fertility that this activity confers. Soil nitrogen turnover—the nitrogen-supplying power of the soil—is lower whenever microbial populations are diminished, and there are consequently fewer invertebrates such as earthworms, ground-dwelling beetles, and nonparasitic nematodes.

Soil systems that are low in SOM and soil biota, whether for either management or for natural reasons, will be lower in fertility, and for this reason they require substantial external inputs to maintain crop productivity. Restoration and maintenance of SOM in both residual and living forms is thus a crucially important management challenge.

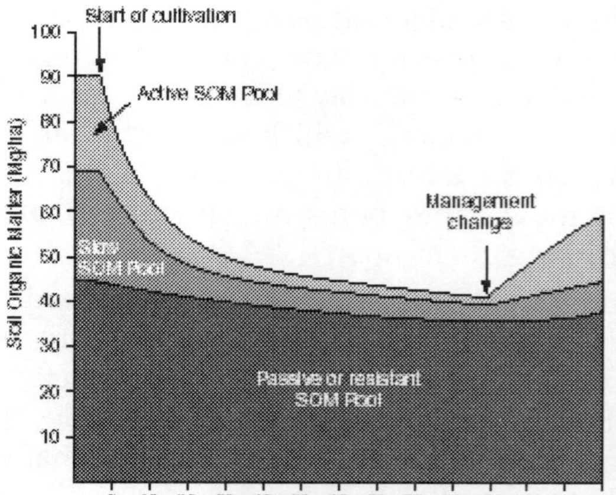

Figure 1. Changes in soil organic matter fractions following cultivation of a soil profile under native vegetation.

Nutrient Availability

Most cropping systems use and export nutrients at prodigious rates. Some nutrient loss, such as that exported in yield, is unavoidable. Other losses, however, such as nutrients lost via hydrologic and gaseous pathways, are inadvertent. All exported nutrients that can limit crop performance must be replaced for a cropping system to remain productive, from external sources or from within the soil system.

Maintaining this nutrient availability in both time and place to match plant needs is one of the toughest of agronomic challenges. For certain plant nutrients such as calcium and magnesium, most temperate-region soils can maintain a steady supply with little depletion even in the face of significant export. This is because the mineral stores of these nutrients are high in most young soils. For other nutrients, however, particularly nitrogen, phosphorus, and potassium (N, P, and K), the ability of a soil to fully resupply losses is eventually lost. When this occurs, modern cropping systems rely on fertilisers to make up the difference.

Nitrogen deficits are especially severe because nitrogen losses can occur via so many different pathways, and it is nitrogen that typically limits the productivity of even natural ecosystems that are not harvested. The two main strategies for improving nutrient availability in cropped ecosystems are to increase inputs and to reduce losses. Inputs are commonly increased via organic or synthetic fertiliser additions, or specifically for nitrogen, by N_2 fixation.

Losses can be reduced, on the other hand, by increasing system-wide nutrient-use efficiency. Nitrogen is a case in point. A highly productive maize crop with a yield of 10 tons of grain removes about 260 kg N ha^{-1} or around 5.2 tons of nitrogen over 20 years of cropping. In uncultivated arable soils, organic nitrogen stores can be as high as 10 tons of nitrogen ha^{-1} on average. Continuous cropping of maize thus has the potential to remove, within 20 years, an amount of nitrogen equivalent to 50% of the nitrogen stock in the native SOM, demonstrating the potential for rapid soil nitrogen depletion. Because nitrogen is the most common limiting nutrient in temperate region ecosystems, restoring lost nitrogen is a crucial agronomic goal.

Preventing as much nitrogen as possible from inadvertently leaving the system is equally important, fromboth an agronomic and environmental standpoint. Improving a cropping system's nutrient-use efficiency requires matching soil nutrient release—whether from organic or inorganic sources—with the demand for nutrients by plants. This matching has to occur both temporally and spatially. In diverse native plant communities and many cropped perennial systems, soil microbial activity will almost always coincide with periods when there is at least some plant need.

In native communities, the presence of diverse species having different life cycles means that at least some plants will be actively photosynthesizing whenever temperature and moisture permit. In the annual monocultures typical of temperate-region agriculture, on the other hand, such synchrony is rare. Most grain crops, for example, are in the ecosystem for only 90-100 days, and only during 30-40 days at midsummer will they be accumulating biomass at a significant rate. In maize, for example, nitrogen uptake rates can reach the astonishing rate of 4 kg N ha^{-1} day^{-1}. This high rate is sustained for only 3-4 weeks, however, and it falls to nil within the following 2-3 weeks.

The much longer periods during which atmospheric nitrogen deposition occurs and soil temperature and moisture are sufficient to support microbial nitrogen mineralisation do not match crops' peak nutrient demand. This asynchrony creates a huge potential for nutrient loss and for low system-wide nutrient-use efficiency.

Spatial symmetry can be as important as temporal synchrony for ensuring that nutrient availability and uptake are well matched. Row crop management, unfortunately, does not often result in well-matched spatial arrangements of plants and resources within a field, and this mismatch also reduces system-wide nutrient-use efficiency. Row vs. between-row differences in soil-nutrient availability have been recognised for decades, and a number of management strategies, discussed later, can be derived from knowledge about how to increase the water and nutrient-use efficiencies of row crops.

Spatial heterogeneity at larger scales is also emerging as a management issue. Available evidence suggests that soil nitrogen availability is highly variable in natural communities, with variable patches of soil fertility at scales that can affect individual plants. High nutrient-use efficiency from both the spatial and temporal perspectives is thus an important goal of agronomists, and one whose achievement depends on adept combinations of soil and plant management decisions.

Solutions to the Soil System Challenges

Restoration of SOM

Decades of research have demonstrated that SOM can be restored and maintained at relatively high levels in most arable soils. Most importantly, those biologically active SOM fractions most rapidly lost following cultivation—such as light-fraction (LF) or particulate organic matter (POM)—can be regenerated. LF has a rapid turnover time of 2-3 years because it is relatively free of mineral material and humification and has high concentrations of carbon and nitrogen.

LF is thus an ideal source of energy and nutrients for microorganisms, and its decomposition releases plant nutrients to the soil solution. Restoring LF and other active SOM pools through strategic crop and soil management thus has the potential to stabilise cropping systems and reduce dependencies on external inputs. At the simplest level, SOM change is simply the difference between organic carbon added to soil and organic carbon lost via the biological oxidation of SOM carbon to CO_2 carbon. There are thus two ways to build SOM in cropping systems:

- increase soil carbon inputs via crop residues, cover crops, and soil amendments such as compost and manure, and
- decrease soil carbon loss by slowing decomposition and (where important) soil erosion.

Carbon inputs to soil are influenced by nearly every facet of agricultural practice. These include crop type and productivity, the frequency and duration of fallow periods, and fertiliser and residue management. Organic amendments such as manure, compost, and sewage wastes provide additional management interventions. High crop productivity based on associated residue inputs does not in itself guarantee higher SOM pools. Relationships between residue inputs and SOM are complicated by changes in enzyme dynamics and decomposition processes following N-fertilisation and other agricultural practices.

In the U.S.A.'s corn belt, for example, even though aboveground residues in a maize ecosystem may exceed by a factor of two the amount of litterfall in the forest or native prairie that the agricultural system replaced, SOM levels in the maize system persist at about 50% of the levels in native forest even when maize residues are not exported from the soil system. This said, substantial residue inputs are still a prerequisite for building organic matter stores in soil. Removing all aboveground residues—as is the case for corn silage, wheat straw production, or biobased fuel production, for example-removes a major source for SOM accumulation.

With other factors held equal, in fact, field experiments have generally found a close linear relationship between the rates of residue carbon return and the SOM levels found in temperate agricultural soils. Organic amendments also provide a direct and effective means for building SOM. For example, in a long-term continuous wheat experiment at Rothamsted, U.K., plots receiving farmyard manure over a 100-year period effectively doubled their SOM levels. Decomposition rates of crop residues and SOM are principally influenced by climate, by the chemical composition or quality of the residue, and by soil disturbance.

In general, decomposition occurs faster in warmer, moister soils, and with management that exposes the soil's surface to greater solar radiation or that uses spring tillage to accelerate soil drying and warming following a winter snow cover so as to promote decomposition. Draining wetland soils for agriculture achieves essentially the same result. Decomposition is also affected by litter quality. Plant tissues lower in nitrogen and higher in structural compounds such as cellulose, suberin, and lignin decompose more slowly than tissues that are higher in sugars, protein, and nitrogen: for example, soybean leaves decompose much faster than do wheat straw or maize stalks.

Few microbes are able to degrade the complex chemical structure of lignin, whereas simple organic compounds can be respired by most soil organisms. It follows that SOM is likely to accumulate faster with the addition of more structurally complex materials, although these relationships may be complicated by interactions between decomposition products and soil physical processes. For example, the rapid production of polysaccharides associated with the decomposition of legumes can facilitate aggregate formation and increased physical protection of SOM.

More research is needed, however, to determine how plant and microbial communities interact to control decomposition and, in particular, the formation of particular biochemicals which stabilise SOM in agricultural soils. Manure tends to be more complex structurally than are uncomposted crop residues because it has already been exposed to microbial attack in the animal gut. Conservation tillage can also conserve SOM by reducing erosion in landscapes subject to wind and water erosion. Cover crops that maintain plant cover during periods when the primary crop is not present—late fall, winter, and early spring, for example—can also reduce the potential for soil erosion and add additional vegetative residue to the SOM pool.

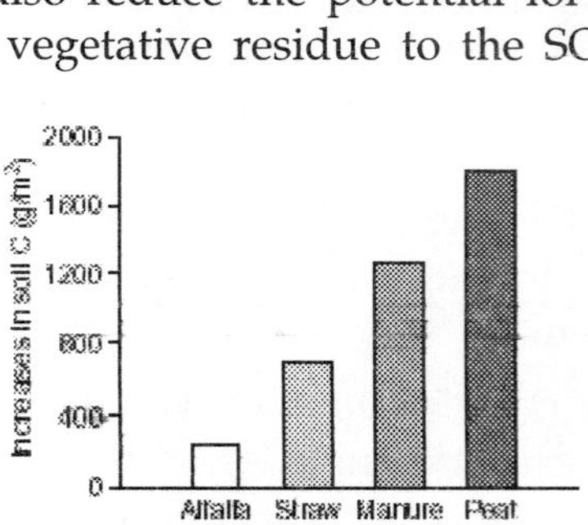

Figure 2. Soil carbon increases over 20 years following the addition of carbon sources differing in a structural complexity or quality to a sandy soil in Canada. Residues were added at the rate of 500 g m^{-2} $year^{-1}$

Restoring SOM in cropping systems can thus be achieved best through some combination of increased organic matter inputs, no-till or other conservation tillage practice, and cover cropping. While any organic matter inputs will help to build SOM, the most effective will be those that are slow to decompose, such as low-nitrogen, high-lignin crop residues or compost and manure. Rotational complexity may also help to restore SOM when one or more crops in the rotation have higher lignin contents, more residue, or a longer growing season than others.

Developing Nutrient Efficiency

Making nutrients available mostly when and where they are needed by the crop improves nutrient-use efficiency. Temporal synchrony is achieved by applying inputs as close as possible to the time required for crop growth. Applying mobile fertilisers, such as nitrogen, in split applications, e.g., 20% at planting and the rest just before the period of greatest crop growth, is common in many temperate systems, although—egregiously -fall application of anhydrous ammonia is still common for maize production in some regions of the Midwestern U.S.A.

Likewise, encouraging decomposition of the previous crop's residue early in a crop's growing season is also beneficial. Spring rather than fall tillage will keep more nutrients in active SOM pools where they are better protected from overwinter leaching and gaseous losses and will serve to stimulate decomposition prior to crop growth. Rotary hoeing or some other type of shallow cultivation well into the growing season can stimulate microbial activity just prior to major crop growth.

Winter cover crops—particularly fast-decomposing high-nitrogen crops such as legumes—also help to provide active-fraction SOM when the crop most needs it. Cover crops can additionally help to capture nutrients released to the soil solution when the main crop is not present; plants active in the fall and spring when microbes are actively oxidizing SOM can temporarily immobilise nutrients that would otherwise be vulnerable to overwinter or springtime losses from the ecosystem. Other aspects of crop management that may influence microbial communities and decomposition are nitrogen fertilisation, inputs of labile carbon compounds, and irrigation.

Many studies have demonstrated that nitrogen or organic matter additions may result in a change in the mineralisation of native SOM. This is referred to as the priming effect: a strong change in the turnover of SOM in response to a soil amendment. Priming effects may play a critical role in controlling carbon balance and nitrogen turnover in ecosystems. However, our ability to exploit the underlying microbial processes to manage soil fertility is currently limited. This is primarily because environmental controls over priming responses are very complex and include interactions between nutrient availability, litter quality, soil texture, and other factors.

Despite these challenges, this should remain an area of active research because the potential benefits are great from being able to manipulate SOM turnover and nutrient mineralisation when and where it is most needed with relatively modest additions of nitrogen or carbon to the soil. Spatial synchrony or coincidence can be achieved at two levels. At the row vs. between row level, inputs such as fertilisers can be applied in bands next to or over the tops of rows using drip irrigation, fertiliser banding, or foliar feeding; or organic amendments or crop residues can be mounded into rows using techniques such as ridge tillage.

Ridge tillage, a popular soil management technique for many low-input farmers in the Midwestern U.S.A, minimises spatial asymmetry by periodicallymounding the between-row Ap horizon into semi-permanent ridges on to which the crop is planted. This concentrates the labile organic matter and soil biotic activity within rows, achieving the same effect as fertiliser banding. At the larger field scale, variability can be addressed by using site-specific application technologies. Many harvest-combines today are sold with global positioning system (GPS) equipment to permit highly-resolved yield mapping.

With proper application equipment, these maps can then be used to tailor fertiliser applications to the productive capacity of any given area of the mapped field. Rather than fertilising an entire field with a single, high rate of application, the highest rates can be applied only where productivity, and therefore plant nutrient uptake, will be high, reducing nutrient losses from lowproductivity areas. In effect, this method uses plants in the field as bioassays for the nitrogen made available by soil microbes; it provides additional nitrogen fertiliser in proportion to the plants' abilities to take it up.

In most temperate regions, the current cost of fertiliser is low relative to the marginal increase in productivity that can be gained from applying it at high rates. So from the producer's standpoint, it rarely pays to reduce the inputs of limiting nutrients that are inexpensive, e.g., nitrogen. Thus, socioeconomic influences condition decisions about achieving crop nutrient-use efficiency. In many if not most cases we cannot expect improved ecosystem nutrient-use efficiency until policy and other issues affecting farmer decision-making are appropriately resolved.

References

Buol, S. W.; Hole, F. D. and McCracken, R. J., *Soil Genesis and Classification*, First, Ames, IA: Iowa State University Press, 1973.

Jones, J. A. A., "Soil piping and stream channel initiation", *Water Resources Research* 7 (3.): 602 - 610, 1976.

Nikiforoff C. C., "Reappraisal of the soil: Pedogenesis consists of transactions in matter and energy between the soil and its surroundings", *Science* 129: 186-196.

Smith, C. J.; Peoples, M. B. Keerthisinghe, G. James, T. R., "Effect of surface applications of lime, gypsum and phosphogypsum on the alleviating of surface and subsurface acidity in a soil under pasture", *Australian Journal of Soil Research*, 1994, 32.

6

Managing Land Productivity

Subhumid to semiarid areas are characterised by rainless periods, both within and between rainy seasons, which are generally unpredictable. Because of this the output of crops, pastures and streamflow is affected not only by the total amount of rainfall in a particular season, but also by the frequency, duration and severity of water stress in the plants at different stages of growth.

Greater attention to the value, capture and use of rainwater in increasing production from rainfed lands in the tropics and subtropics is justified on two main counts:

- Increasing numbers of the rural poor live in areas where they must depend on rainfall alone for both crop production and domestic water needs.
- Since yields of crops in small farmers' fields are on average far below those possible on well-managed plots on experimental stations even modest actual increases in production are probably achievable, whereas proportionately larger increases from irrigated areas seem unlikely to be achieved.

Soil productivity should be maintained and improved overtime. Two features are fundamental, for without them plant growth will be limited and the productivity of soils will not be sustainable:

- Sufficient soil water, in optimum proportions with pore space and solids, and of sufficiently long persistence at plant-available tensions, is vital for plants to complete their full sequence of growth.

— For damaged soils, achieving and maintaining optimum porosity, plus raising and maintaining biological self-recuperation capacity, are effective ways of improving crop production where rainwater is a limiting factor.

Successful water management in a dryland farming system is based on: (1) retaining precipitation on the land; (2) reducing evaporation; (3) utilising crops that have drought tolerance and fit rainfall patterns. This raises three questions:

— Can water get into the soil fast enough to avoid runoff?

— Is the soil in a condition to allow water uptake through plants without their suffering damaging water stress in their tissues and to allow downwards transmission of excess to the groundwater?

— How is it possible to raise people's skills in soil and crop management to close the yield gaps between research-station experience and in-field practice?

To effectively address the rising concerns about the land's capacity to produce crops and sustain streamflow, it is no longer sufficient to consider macroscopic factors alone. A framework for action must be based on understanding at microlevel as well. This will include understanding of how plants and soils function together and how they are likely to react to proposed improvements, e.g.:

— The collapse or compaction of pores of all sizes is the prime reason why water may not enter the soil and runoff can occur.

— A key factor for soil sustainability is the maintenance of biological capacity for self-recuperation and how to encourage this biological activity in the field.

More widespread understanding of such factors may lead to a greater respect for the soil as an environment for biological activity, for meso- and micro-organisms as much as for roots themselves.

Deterioration of Water Supply

A deterioration of water supplies refers to diminished quantities of groundwater and surface water as well as to deteriorating water quality. Poor quality water may be the result not only of inappropriate land use and soil management practices which result in materials being transported by surface runoff, but also of industrial and urban pollution due to inadequate processing controls and poor sanitation. Increased runoff at the expense of rainfall infiltration is a major cause of declining groundwater, as less water is then available to percolate through the soil down to the groundwater, i.e. less recharge occurs.

Increased runoff is often the result of changes in land use that reduce the protective ground cover and decrease surface soil porosity, as for example when forest vegetation is converted into inadequately-managed annual cropping. Such land use changes often arise when rising population pressures force people to cultivate or graze land that is poorly suited to the use to which it is being put. Changes in land use that increase the quantity of water used in transpiration, such as reforestation programmes, will be expected to diminish the frequency and amount of groundwater recharge, assuming no changes in the amount of rainwater lost by runoff or other processes.

Conversely, deforestation followed by the cultivation of annual crops would be expected to decrease transpiration and so increase groundwater supplies, as long as no extra runoff occurs as a result. Drainage of swampy areas in middle and upper watershed positions can also reduce the amount of water reaching the groundwater through deep drainage due to the diversion of water into drainage canals. Falling groundwater levels may arise as a result of increased water consumption by irrigation schemes.

The lack of proper drainage in irrigation schemes may lead to deteriorating groundwater quality due to the accumulation of salts. Greater runoff can also result from urbanization because of the replacement of agricultural land by extensive areas of tarmac and concrete, such as roads, pavements and buildings. These prevent water infiltration and generate high proportions of runoff. In many developing countries, as populations grow and industrialisation and urbanisation increase, the demand for water grows and eventually exceeds the quantities available.

Throughout the world's continents water tables are falling, and it has been estimated that by 2025 more than half the world's population will be living in regions suffering from a shortage of water. The combination of falling groundwater and greater runoff will reduce the base flows of rivers and streams, and will greatly increase peak flows and the incidence of floods. High runoff often affects the quality of surface water by its load of eroded soil sediments, may make the water unsuitable for drinking, and may increase the costs of water treatment.

High sediment loads in the reservoirs of hydroelectric schemes will reduce the life span of dam sites and increase turbine maintenance costs. The recommended amount of water needed for one person per day for cooking, drinking and washing is about 50 litres, but the amount of water needed every day for a crop to transpire and produce sufficient grain for one person is some 10 to 20 times larger. Therefore, water shortages will have most effect on food production, rather than on the availability of water for domestic use.

Indicators of Deterioration

The following are simple visual indicators of deteriorating quantity and quality of water supplies.

Indicators of the lowering of groundwater tables from 1 year to another:

— drying of wells
— drying of springs
— extending the depth of boreholes
— dying trees at river margins

Indicators of reduced surface water:

— diminished base flows in rivers
— increased deposition of sediments in river beds
— more meandering streamflows in river beds
— greater frequency and severity of flooding
— greater deposition of large rocks and boulders

Indicators of reduced quality of surface water:

— pollution and discolouration of water by sediment
— algae
— bad smells

Indicators of reduced quality of groundwater:

— high salt contents
— bad smells
— algae

Soil Productivity and Erosion

Increased possibilities for safe and sustainable intensification of production can be identified if the nature of soil productivity and the process and the effects of soil erosion are examined.

Soil Productivity

Fertility is the inherent capacity of a soil to supply nutrients in adequate amounts and suitable proportions, whereas soil productivity is a wider term referring to the ability of a soil to yield crops. The chief factors in soil productivity are soil organic matter, soil texture, structure, depth, nutrient content, water-storage capacity, reaction and absence of toxic elements. A brief description would indicate that soil productivity depends on physical, hydric, chemical and biologic characteristics and their interaction.

Much can be known about the above-ground growth and development of plants by observing and measuring them and their functions. We know much less of what goes on beneath the surface in the soil ecosystem, where plants' roots are important constituents. In order to function unimpeded through cycles of growth to maturity, land plants require water to pass through them from soil to atmosphere.

The smaller roots and their root hairs also constitute a large part of any root system, but are not easily visible. Root systems are of astonishing dimensions, as illustrated by comparative figures from the same-sized samples of soil. In favourable conditions, roots of some plant species may grow by as much as 10 mm/day. The relative proportions of solids, liquid and gases in the rooting environment are as important as the manner in which they are arranged. In this respect, the biotic content of soil is also important because together with plant roots it contributes to restructuring soil and the improvement of porosity after damage by compaction, pulverisation or structural collapse.

Many degraded soil situations have arisen because of mould-board or disk plough-based farming practices, which have resulted in:

— decline in soil organic matter

— compaction of the soil causing reduced porosity

— reduced plant-available soil moisture at critical times

— loss of soil depth

Each of these four factors negatively affects soil as a habitat for plant roots. The view that such conventional tillage methods limit the development of an optimum habitat for rooting is borne out by the experience of alternative agriculture systems. The amount of moisture in the soil depends on how much rainfalls and enters the soil. Under rainfed agriculture, the amount of water entering the soil depends on what percentage is diverted above the surface as runoff. It may not always be possible to prevent all runoff, but improvement of soil physical conditions will help to reduce it to the unavoidable minimum.

A thin surface crust or subsurface compaction can be enough to reduce the rainwater infiltration rate, to provoke runoff and to cause loss of water, soil and consequently potential soil moisture. However, when the soil is covered with litter, more water enters and the soil surface is protected from the force of raindrops. Infiltration and availability of water to plants depends on how the water is held between, firstly, individual particles of the soil and secondly, on the distribution of different sizes of pore space in soil. If the majority of pore spaces is very small, whether because of the inherent properties of the soil such as in a very clayey Vertisol, or because the soil has been compacted, the water may be held so tightly that plants cannot extract much of it.

On the other hand if the soil has a predominance of large spaces, such as in coarse-textured sandy soils, much or all of the rainfall may enter the soil and pass easily and rapidly down through the profile without much of it being retained. A wide range of pore sizes is therefore desirable for enabling both retention and transmission of rainwater. Reduction of pore space may be at least as important as losing soil particles with respect to yield. It affects water movement and the soil's tenacity of water retention, root expansion and gas exchange of O_2 and CO_2 with the atmosphere. Its loss is similar to losing spaces in a block of apartments when they are demolished: the same quantity of materials remains, but the value of the architecture is lost because there are no longer usable voids/ rooms.

Field observations in Malawi, Zambia and Tanzania indicate that repeated tillage to the same depth with hand hoes can cause subsurface pans of compacted soil at the base of the tillage layer. After a few years, they may become so dense that neither roots nor water can penetrate them easily. This increases surface runoff, severely limits soil depth and causes stunting of roots. With root access to soil moisture restricted to the shallow soil layer above the compacted layer, plants are prone to suffer water stress after only a few days of dry weather. Much of the blame for this damage to soils can be attributed to inappropriate tillage practices. Such problems are found not only in tropical areas but also are now widely seen as well in temperate zones across North and South America, Europe, Asia, Australia and New Zealand.

In temperate zones damage is due to compaction by machines. In tropical regions much damage is also done by high-intensity rainfall on unprotected soils. Physical degradation of the soil is the precursor of excessive runoff, reduced soil moisture and root restriction, and is a primary limitation to crop growth. There are already widespread and serious problems in soils as rooting environments, characterised not so much by erosion as by unexpectedly poor performance of crops in large parts of the world, in both rainfed and irrigated areas.

Soil Erosion

Runoff and erosion occur because soil porosity has been damaged, whether at or below the surface of the soil. They are the consequences, not primary causes of land degradation. In many instances tillage aimed at loosening the soil to let in more rainwater can also result in soil collapse, which then leads to increased erosion and loss of potential soil moisture through runoff.

It is sometimes assumed that yield reductions following soil erosion can be directly related to the quantities of soil materials lost. This may not always be the case, for example where erosion removes a similar quantity and quality of material from three soils with different subsoils, yields may be different from each other and equal, lower or higher than before the erosion. Where a subsurface layer is of better quality for rooting than that which overlays it, erosion could be followed by higher, not lower, yields though such situations are not common.

The differences in yields are related to the differences in the characteristics of the subsurface habitats in which the roots will grow before and after the erosion i.e. differences in depth, organic matter content, infiltration capacity, plant nutrient supply, biotic activity and architectural stability. Differences in soil surface porosity affect water infiltration rates. Compaction of the soil by trampling or machinery has the same effect, changing the hydraulic conditions of the soil. Loss of porosity in the soil increases surface runoff, increasing infiltration increases soil moisture. A failure to understand this relationship has often led to inappropriate actions to stop erosion, such as construction of physical works or overuse of fertilizers.

Conventional physical means of Soil and Water Conservation (SWC) have often proved less than satisfactory and not widely acceptable to farmers, because they tried to halt runoff and erosion rather than concentrating first on improving the absorptive capacity and productivity of the soil in situ, thereby minimizing runoff and erosion as a consequence. In undisturbed conditions under native vegetation, the effective rooting depth for this soil is more than 3 metres. Breaking the pan to restore favourable conditions for rooting and water infiltration would have been more appropriate than using sandbags.

Each time erosion occurs, the rooting environment for the subsequent crop is altered. This understanding shows the need to:

- protect the soil surface from damaging forces of rain and wind;
- increase rainwater infiltration;
- encourage biological restoration of newly exposed surface layers;
- minimize desiccation, which damages both roots and micro-organisms.

This approach is radically different from and should precede any physical measures that may still be necessary to catch and redirect runoff once it has begun. Soil characteristics, which favour water infiltration and gas exchange are the same as those that minimize runoff and erosion. In this way conservation concerns can be fully integrated with the production process.

Drought and Plant-damaging

The figure for the annual average rainfall is no indicator of the frequency of drought, either between or within years, as shown in Table 1. The mean rainfall between 1956 and 1977 was 1 025 mm, but the variation was from 507 to 1 917 mm.

Table 1. Annual rainfall totals at Indore, India

Year	Rain mm	Year	Rain mm
1956	808	1967	964
1957	632	1968	903
1958	1 208	1969	1 054
1959	1 743	1970	910
1960	860	1971	926
1961	1 246	1972	598
1962	1 103	1973	1 917
1963	998	1974	952
1964	1 084	1975	1 209
1965	507	1976	1 221
1966	696	1977	1 025

Drought periods within a particular year may show up as a delay in the onset of a rainy season; as dry spells of a week or more at critical periods of crop growth within the season; or as an earlier-than-expected end of the rainy season.

Only rainwater that enters the soil can be effective with respect to plant growth and dry season streamflow. Avoidable surface runoff can reduce soil moisture and groundwater. Induced drought means that plants may become stressed earlier than need be, even though there is sufficient rainfall above-ground to provide for the crop. A recent study in Karnataka, India shows that a precarious situation develops when the combined demands for soil moisture for plants and liquid water exceed average recharge of soil and groundwater.

Duration of Drought

Climatic drought is unavoidable. Extending the period in which soil moisture remains available to plants shortens the duration of potentially damaging water stress in plants. At the same time this shortens the length of the non-producing period of the year during which the stored food will be eaten before the next harvest. In seasonally dry regions the focus of attention should be on how much rainwater can be caught and stored in the soil much more than on emphasizing how much runoff has occurred across the land surface.

Root systems are more extensive when water is not a limiting factor, as illustrated by differences in root growth by plants of the same clone of tea grown under rainfed and irrigated conditions. After 9 months, the root systems of one representative plant from each treatment were exposed by root washing and drawn to scale on paper. Dry periods within the 1968-1969 rainy season led to inhibition of root growth in the rainfed treatment, whereas the regular provision of sufficient irrigation water avoided water stress in the other treatment and resulted in a more profuse root system. Roots grow and extend within those volumes of soil where soil moisture is available.

This indicates that enabling more rainwater to enter the soil and minimising losses from runoff and evaporation from the surface will be beneficial for root growth, provided that other factors such as nutrient levels and physical barriers to root growth are not limiting. The development of a wide range of stable pore spaces cannot be achieved by mechanical tillage and can only result from soil biological activity. Soil organisms make a major contribution in developing and maintaining porosity and may allow plant survival even after water stress may have caused active growth to cease.

Considered in this way, the severity of climatic drought can be diminished, inasmuch as plant persistence can be extended and the possibility of post-drought recovery is increased. The provision of permeable soil cover, preferably crop residues, moderates high temperatures in the upper root zone. Soil cover also prevents rain splash and encourages infiltration and markedly reduces the rate of evaporation of water from the upper layer of the soil. This conserves moisture, delaying the onset and shortening the duration of severe stress.

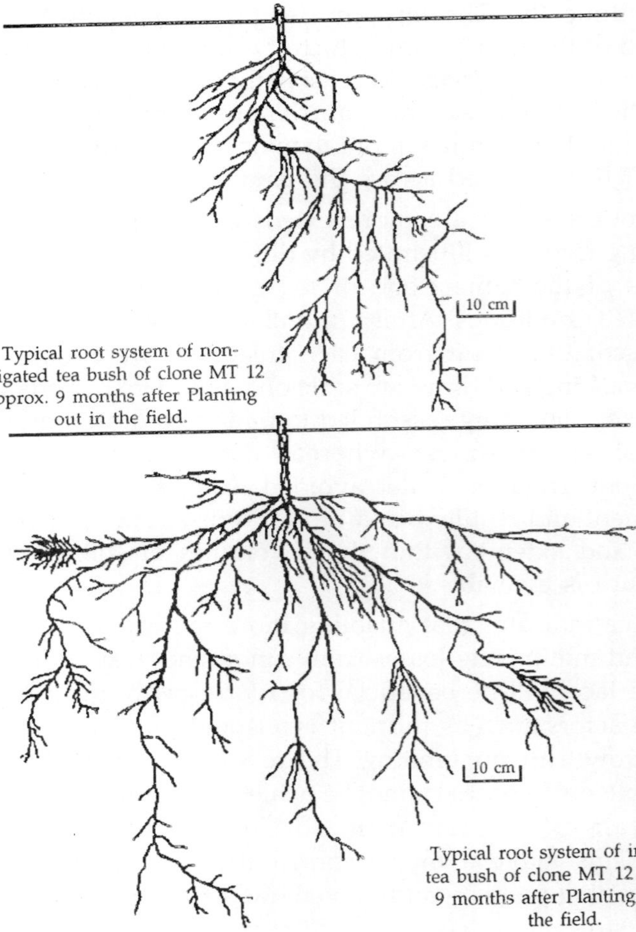

Figure 1. Root systems of two young tea plants of the same clone (MT12) without and with irrigation, after 9 months in the field

Even where little residue cover is available, soils in good condition under minimal tillage may provide better conditions for seedling growth and survival than those damaged by inappropriate heavy cultivation. For instance, in the dry wheat-growing lands of Western Australia, it was stated that in 2000: "...even no-tilled crops suffered severely with drought. However, their revival was markedly better than in situations where the soil structure had been damaged by tillage. Tilled soil did not receive the rain as well as soil that had softened through the years of no-tillage... The crusts from tilled soils are in strong contrast to the soft furrows [made by press-wheels at drilling-time] in the paddocks with a history of no-till.

Changing the Perspective

From this different viewpoint, important changes in emphasis include:

- — Focus on saving pore spaces, more than saving solid particles.
- — Emphasize increasing infiltration more than reducing runoff.
- — To minimise erosion, maintaining a cover of plant residues on the soil is a better first action than building cross-slope banks at intervals downslope.
- — On seeing a muddy river in flood, it is more sensible to ask "why so much water?" than to exclaim at the sediment being transported.
- — Consider Water and Soil Conservation (WSC) rather than Soil and Water Conservation (SWC).
- — Think about the principles of what is to be achieved first, before choosing specific practices.
- — Reduce risks of failure due to drought, rather than bemoaning increased severity of drought.
- — Build soil from the surface downwards, particularly by favouring biotic activity, rather than merely waiting for it to deepen from the bottom upwards.

Care about Roots, Soil Organisms and Water

The way the soil is managed as an environment for roots influences the onset, duration and severity of drought, since the roots are suppliers of water and nutrients to the other parts of the plant. Without sufficient water to satisfy plants yields may be limited after even a few days in hot weather. The more severe and prolonged the dry period, the greater the damage to final yields. A good understanding of the below-ground environment and of how this ecosystem functions is necessary so that it can be managed more appropriately.

The organisms in the soil ecosystem break down and transform organic materials, and contribute to:

- the soil's porosity through burrowing and the formation of aggregate-binding gums;
- the soil's water holding capacity via that porosity and water retention by humus;
- the soil's capacity to retain and slowly release plant nutrients;
- the fixation of atmospheric nitrogen.

Soil organisms are subsurface workers, who perform many soil improving activities without cost to the farmers. They deserve more attention than they generally receive on how best to provide for their requirements. Rainwater for plants' needs should be retained by saving it in the soil where it may benefit all inhabitants of the root zone. Water in excess of these requirements should be able to pass further downwards to contribute to the groundwater, available for use downstream.

These three 'care' suggestions propose an approach aimed at determining the needs of the soil ecosystem of which plant roots are part. This is a necessary step in deciding ways to increase land use intensity without damage to the basic natural resources of water and soil.

References

Coleman, D.C. and Crossley, D.A. Jr., *Fundamentals of Soil Ecology*, 2nd ed., Academic Press, Burlington, MA, 2003.

Kiem, R. and Ko¨gel-Knaber, I., "Contribution of lignin and polysaccharides to the refractory carbon pool in C-depleted arable soils," *Soil Biol. Biochem.*, 35, 101–118, 2003.

Rasmussen, P.E. *et al.*, "Crop residue influences on soil carbon and nitrogen in a wheat-fallow system," *Soil Sci. Soc. Am. J.*, 44, 596–600, 1980.

7

Minimising Water Stress and Improving Water Resources

The porosity of surface soil may have been reduced by clogging of pores with particles detached from soil aggregates under the impact of raindrops, or by the deposition of detached particles on the soil surface as impermeable crusts or seals. The porosity of subsurface soil may be naturally low, or may have been reduced by compaction and tillage practices that have disrupted or destroyed pore spaces causing a zone of low permeability at the base of the tilled layer.

The degree to which soil porosity is reduced by tillage is frequently sufficient to limit root penetration, but is less often so severe that permeability to water is significantly diminished. The overriding approach should be to instil in society, and in farmers, extensionists and researchers in particular, the will to create and sustain soil conditions that encourage the infiltration of rainfall where it falls, and to counteract the causes of runoff. This implies that the porosity of the soil must be at least maintained, or increased.

Porosity of the Soil Surface

Porosity of the soil surface is best maintained by first protecting it from the disruptive action of raindrops through a protective cover, usually of residues from the previous crop, a cover crop or mulch, and by ensuring the soil is not disturbed by tillage. The effects of conservation agriculture on higher infiltration and reduced runoff and flooding have been well documented in Brazil in particular.

If the whole concept cannot be applied immediately, improvements in soil moisture status of the soil can still be achieved, though probably not to the same extent, by other measures aimed at prolonging the useful life of rainwater. These include the use of surface residue covers alone, fallow periods under cover crops or natural vegetation, protection or temporary closure of grazing lands and forests from overgrasing, and operations on the contour, complemented by physical measures to detain rainwater. The regular use of shallow tillage with disc or tined implements to break-up surface crusts to increase surface porosity and enhance rainfall infiltration is not recommended.

The increase in surface porosity is only temporary and on crusting-susceptible soils tillage will need to be repeated after every rainstorm. Tillage leads to the disruption of pore spaces in the soil, and the use of discs, in particular, often causes compaction, which may impede root growth and rainwater percolation. Tillage also accelerates the loss of soil organic matter leading to a progressive deterioration of soil architecture and a reduction in the number and stability of pores that allow growth of roots and movement of rainwater. Regular tillage therefore is not recommended as a solution to restricted infiltration caused by low porosity of the soil surface.

Surface Residue Covers

A residue cover absorbs most of the energy of the raindrops that fall on it and by the time this rainwater reaches the soil below, its ability to disintegrate soil aggregates and detach fine particles is greatly reduced. Consequently, there is little or no clogging of surface soil pores by detached particles, and little deposition of soil particles that would form a crust on the surface.

The benefits of a residue cover are most apparent on soils initially in reasonable physical condition, but even under these conditions runoff can sometimes occur despite a good soil cover. For example, runoff will occur when rainfall intensity is greater than the soil's infiltration rate, or when the soil's pore spaces are already filled with water because the soil is shallow, its water holding capacity is low, or its subsoil is only slowly permeable.

When a residue cover is applied to a soil with a very degraded surface of low porosity, the beneficial effect of the cover on infiltration may be initially limited. In such situations, it is advisable to accelerate the recuperation of surface porosity before applying residue covers by tilling the soil once to break-up the crust and any subsurface pans, followed by a fallow period under a cover crop to enhance the formation and stabilisation of soil porosity.

The choice of a cover material depends on what is locally available. Residue covers may consist of:

- Crop residues left in the field after harvesting the previous crop.
- Cover crops sown the previous season and left on the soil surface after slashing or applying a herbicide.
- Leaves and branches lopped from trees growing within the cropping area.
- Mulches of grasses, shrubs, weeds, litter, husks and other organic waste materials.

The last option (mulches) requires residues to be collected from elsewhere, transported to the cropping area and then applied in the field, whereas in the other options, the residues are produced within the cropping area. Examples of materials that may be used as mulches are grasses and sedges, banana leaves and pseudostems, shrubs such as Lantana and wild sunflower (Tithonia), forest litter and tree loppings. Other materials occasionally used are weeds, rotten thatch and coffee husks.

Where soils have a cover of stones, these may be left on the surface as a protective cover provided they do not interfere with planting or weeding operations. Mulching is most commonly practised on horticultural crops that produce negligible residues (foliage), or are completely harvested for their foliage, or are completely harvested.

In the steeply sloping Guaymango area of El Salvador, efforts were made in the 1960s and 1970s to improve crop production by encouraging small farmers to adopt a combination of hybrid seeds, nitrogenous and phosphatic fertilisers, increased plant densities and application of herbicides and insecticides to maize, sorghum, sesame, rice, beans. These recommendations were not particularly successful and in 1973 recommendations for soil conservation were added. These included no burning of crop residues; uniform distribution of residues across the field; use of living or dead barriers; and sowing on the contour in a zero tillage system.

Improvements in crop yield and quality of soil occurred and a high proportion of farmers adopted these measures. Although erosion control was cited as the farmers' main reason for not burning crop residues any more, an important pointer to the benefits due to improved soil moisture conditions was evident in 1997. In that year there was a serious drought during the rainy season associated with the El Niño weather phenomenon.

But according to the farmers, they were able to harvest almost as much maize as in a more normal year because of conservation of moisture in the soil as a consequence of the better soil status, while neighbours who had not adopted the system lost their crops to drought. Nor did they lose their crops the following year during hurricane Mitch and the associated torrential rainfall, which caused disastrous flooding.

The farmers noticed that the same mulch prevented the seeds from being washed away by rainstorms and facilitated rainwater infiltration so that they did not have problems of decaying plants during the heavy rains. A cross-check on this beneficial effect under the same extreme weather conditions comes from Honduras, where hurricane Mitch caused much erosive devastation on many hillsides, but less on those hillsides where soils were well protected by crop residues. In a limited area of western Honduras the Quesungual traditional agroforestry system has been used by small farmers to produce maize, sorghum and beans.

As rising population pressure makes the traditional slash-and-burn system increasingly unsustainable, there is increasing interest among farmers in the Quesungual system. It combines pruning of naturally regenerating indigenous trees and shrubs with normal agroforestry methods for growing high-value timber and fruit-trees. Before sowing, vegetation is cut down by hand without burning and is spread across the field together with the branches and leaves from pollarding. Crop seeds are then scattered or jab-planted with a stick through the mulch layer. Weeds are cleared infrequently, by hand or using a herbicide.

Surface Residue Covers and Rainwater Infiltration

The physical contacts between a residue cover and the soil surface obstruct the movement of the runoff, slowing it down, giving more time for infiltration and so reducing the volume of runoff. Thus two aspects of surface cover can be distinguished:

— all surface cover absorbs the energy of raindrops and so prevents the loss of pore spaces into which rainwater can infiltrate;

— contact cover slows down any runoff, giving more time for infiltration.

The degree of contact cover is important especially on steep slopes, on soils with naturally low infiltration rates, and on degraded soils with surface crusts or seals of low porosity. Furthermore, it is the contact cover that is immediately accessible to soil macro-organisms and can stimulate their activity. Thus greater numbers of biopores are likely to be formed, leading to more rapid infiltration and percolation. This is why major disturbances such as tillage or incorporation of residues, mulches or other organic matter drastically reduces these positive effects.

Pliable materials of short length, such as leaf or grass mulch, which can be easily flattened by raindrops, will develop a high degree of contact cover and will substantially slow down the speed of runoff flow, generally resulting in reduced volumes of runoff. In contrast, inflexible long materials, such as woody branches of tall bushes that are not easily flattened by raindrops, will develop a low contact cover and so have less influence on the speed of runoff flow.

Advantages

The advantages of mulches are the same as for crop residue, i.e. increased infiltration, decreased runoff, and greater soil water availability. They both provide additional benefits, notably less soil water losses by evaporation, less weed incidence and water losses by transpiration, softer and more workable soils, increased earthworm activity, the incorporation of additional nutrients and frequently increased yields.

In western Kenya, mulching with Tithonia has given substantial yield increases of maize, kale, tomatoes and French beans. Net profits from mulching kale ranged from US$91 to US$1 665 per ha. In the semiarid zone west and north-west of Mount Kenya, maize yields increased by a factor of 4.4 when 3 t/ha of mulch were applied. Termites were not a problem in this area, probably because of the cool climate.

Disadvantage

The main disadvantage of applying mulch is the cost or labour of collecting, transporting and applying the mulch. This is not the case with crop residues, which are produced on-site. Often, there will be no suitable mulching materials in the vicinity of the farm, or there is insufficient labour available. Transporting large quantities of mulch for large-scale cropping is seldom economic and mulches cannot be applied after emergence to closely spaced crops.

When a cover crop is used as mulch, there is the cost of slashing the cover crop or applying a herbicide. Similarly, lopping trees and distributing the branches and leaves over the cropping area requires considerable labour. On steep slopes, the application of residue covers is not easy and requires much labour as well. Moreover, these materials are easily washed downhill on steep slopes.

Mulching materials and crop residues are often grazed by cattle belonging to the farmer, the community or the landowner (in the case of tenant farmers), fed to livestock, or sold as fodder. Sometimes these materials are in demand for thatching or fuel; in many semiarid areas they are rapidly consumed by termites, and in hot humid climates, they decompose rapidly.

Another disadvantage of mulches is a progressive decrease in soil fertility where the mulching materials are produced, unless manures or fertilizers are applied. In parts of Uganda, the residues of cereals grown on hillsides are used to mulch bananas on the lower slopes or valley bottoms, which become enriched in nutrients at the expense of the cereal areas. Soil erosion may also degrade the source areas when the cover provided by the vegetation is removed for use as mulch.

Amount of Residues

In relation to increasing infiltration, studies over two seasons in Nigeria on slopes of 1 to 15 percent have shown that 4 t/ha of rice straw mulch, equivalent to about 80 percent cover, will reduce runoff to 5 percent of the seasonal rainfall. A similar result has been found on a 12 percent slope with a well-structured freshly cultivated soil in Kenya, where 4 t/ha of grass mulch equivalent to 79 percent cover, reduced the runoff from simulated rainfall to 5 percent. On the basis of these data an 80 percent cover, equivalent to about 4 t/ha maize straw, would appear to be appropriate for increasing rainwater infiltration.

Favouring Conditions

The use of soil covers is more common in subhumid and humid zones because of the greater availability of vegetative materials. Nevertheless, they are particularly suited to semiarid areas when materials are available and in the absence of severe termite problems. Mulches are often applied to limited areas of high-value horticultural crops and home gardens in easily accessible fields with gentle slopes.

Natural Vegetation

When soils are so badly degraded that they must be taken out of production, soil porosity can be restored through the action of biological processes. This can be achieved by fallowing for 1 or several years under natural vegetation, natural vegetation enriched with fast-growing leguminous trees, or planted fallows. The accumulation of large amounts of biomass on the soil surface from the fallow vegetation associated with high biological activity and strongly developed root systems promote the biological recuperation of soil porosity.

Biological incorporation of residues into the surface soil results in higher soil organic matter in the upper few millimetres, which progressively extends into deeper layers overtime. The permanent cover of surface residues encourages soil faunal activity, which combined with higher soil organic matter contents leads to improved soil porosity.

A well-adapted, deep-rooting leguminous cover crop often speeds up the recuperation of soil porosity compared with a natural vegetation fallow because larger amounts of biomass are rapidly produced by the cover crop. Whereas a natural vegetation fallow may require 3-5 years, a cover crop may recuperate soil porosity in 1 year. When degraded soils are severely compacted, deep tillage with a subsoiler immediately prior to sowing the cover crop encourages establishment and development of the cover crop. If the degraded soil is severely deficient in phosphorus the application of P fertiliser will be necessary to encourage the establishment of the cover crop.

A constraint of soil recuperation by natural vegetation fallows in mechanised production systems is the problem of eliminating trees and excavating roots before returning to cropping. If a manual system is to be adopted, the problem is less serious. Herbaceous and shrubby cover crops can be eliminated much more easily by slashing, mowing or application of a systemic herbicide, and the subsequent crop may be sown directly into the residues of the cover crop.

Temporary Closure of Grasing Lands

Low infiltration and high runoff can occur on grasing lands even on slopes less than 2 percent, as for example at Sebele, in Botswana. In this area, vegetation cover was considered to be the most important factor controlling infiltration and runoff, and catchments with a cover in excess of 70 percent generally had lower runoff compared with those with less than 70 percent cover. Although the percentage of grass cover in grazing lands has an important influence on rainfall infiltration, soil surface porosity can be more important, especially when overgrasing has degraded the soil, resulting in surface compaction and very low porosity.

On degraded grazing lands at Iiuni, Kenya, for example, even with 57 percent vegetative cover the runoff was in excess of 60 percent. The presence of algae growths on bare surfaces that were resistant to wetting encouraged runoff, whereas stone covers reduced runoff due to the creation of water-storage areas between the stones where the rainwater is detained, allowing more time for infiltration.

Forest Protection and Water Infiltration

Forest provides an excellent protective cover made up of the canopy, low-storey bushes, herbs and surface litter, which combine to protect the soil surface from loss of porosity by direct impact of raindrops. The litter also serves as a food and energy source for soil organisms, which encourages the formation of soil organic matter and faunal passages leading to high infiltration rates.

Where forests are not protected from grazing and litter is consumed by livestock, removed for use as mulch as in parts of Nepal, or is lost in fires, the surface cover may be diminished to such an extent that the soil becomes bare. This is likely to be more serious under trees that discourage the growth of understorey herbs and shrubs, such as teak (Tectona grandis) and some species of Eucalyptus due to shade, high water use—especially by Eucalyptus—and to a lesser extent because of the acid nature of the litter.

If the tree canopy is high, accumulated rainwater drops that fall off the leaves may be larger than normal raindrops and can fall with sufficient velocity to cause more damage to the soil than if there were no tree cover. This can lead to a pronounced loss of soil porosity, as can trampling by livestock, resulting in restricted infiltration and high runoff despite the high canopy cover. The protection of forests from overgrazing is an important management issue in overcoming restricted infiltration, and the establishment of forest user groups is often a crucial step in effectively controlling overgrazing and the loss of surface soil porosity. Forest user groups are most likely to be successful where indigenous forest management systems have existed.

Solutions to Restricted Infiltration

Alternative, but less favourable solutions to restricted infiltration are the use of physical structures, which may be necessary under certain situations:

— When it is not immediately feasible to implement conservation agriculture or simple soil cover because, for example, crop residues are used as fodder.

— As backup measures to support conservation agriculture where the problem of restricted infiltration is due to rainfall intensities that are higher than soil infiltration rates even in the presence of a residue cover.

In these situations, the volume of water soaking into the soil may be increased by giving more time for infiltration by slowing down runoff, by means of physical or vegetative structures constructed across the slope and parallel to the contour. Closely spaced structures on the contour may be formed over the whole field so that rainfall is detained where it falls.

Widely spaced structures at intervals down the slope used on their own without contour field operations between them will result in rainwater running downslope until it is detained or slowed down at the next barrier. Details of the layout, design, construction and maintenance of these structures appear in many Soil and Water Conservation (SWC) handbooks, such as Soil conservation, Soil and water conservation manual for Kenya, FAO Soil Bulletin 70, A land husbandry manual and other documents produced by governmental and other agencies for specific countries or particular environmental conditions.

Contour Field Operations

On sloping land all field operations such as tillage, planting, weed control, spraying and harvesting should be carried out along the contour. Ridges and mini-depressions along the contour create small storage volumes where rainwater can accumulate, allowing more time for infiltration. Field operations conducted in a downslope direction can cause a devastating impact resulting in high runoff losses and soil erosion.

Narrowly spaced contour planting ridges with and without cross ties have the advantage of detaining rainwater where it falls so that there is more time for soak-in, and can be an effective means of encouraging infiltration and preventing runoff in semiarid and the drier subhumid areas. An additional advantage is that working along the contour makes operations such as harvesting easier and quicker.

Constraints

The surface depressions have limited capacity to retain water and on sloping land the effective storage volume rapidly diminishes as slope increases. On slopes greater than 5 percent the effective storage volumes are considerably reduced. Reductions in storage volume will also occur on soils with a low structural stability, as the small ridges slump into depressions on becoming wet. Substantial runoff can occur even on land of 1-2 percent slope when the soils are of low stability and susceptible to crusting. Even on structurally stable soils, depressions may be quickly overtopped by the accumulation of rainwater from all but the lightest of rainstorms.

Adoption of Contour Field Operations

The only exceptions to contour cultivation may be in high-rainfall areas where the soils have high infiltration rates and high susceptibilities to mass movements, e.g. landslides and mudflows. In these situations high soil water content increases the risk of mass movements, and so it may be better to encourage controlled runoff of some of the rainfall. Since the effectiveness of contour field operations in reducing runoff is limited on all but the gentlest slopes, it should be considered as just one of the practices necessary to increase water availability.

Contour Planting Ridges and Tied Ridges

In tied ridges the ties are constructed at intervals across the furrows formed by the contour ridges. These structures are usually constructed with animal traction or tractor power and may be formed annually or can be semi-permanent. They may also be made by hand but labour demands are high. The precise form and management of contour ridges and tied ridges vary considerably, with the optimum design and management being dependent on the crop, rainfall and soil type.

Contour ridges run the risk of being overtopped if too much rainwater accumulates within the furrows. They also may be breached or collapse at low points where large volumes of runoff accumulate from along the furrows. If large volumes of water frequently accumulate a subsurface pan or horizon of restricted permeability may be present beneath the furrows. These risks can be reduced by carefully laying out and maintaining the ridges and furrows to ensure there are no low points and by constructing tied ridges to prevent lateral movement of water along the furrows towards any low points that may exist.

The ties should be spaced at 1 to 3 metre intervals along the furrows and no more than half to two-thirds the height of the ridges. Although tied ridges require additional work, they provide good insurance against the collapse of ridges at low points during heavy rains and the loss of rainwater by discharge from the ends of the furrows if a slight gradient exists. The furrows of contour ridges are normally aligned parallel to the contour. However, if very large volumes of runoff are periodically expected, tied ridges should be installed and the furrows constructed on a slight gradient so that excess rainwater is discharged along the furrows to prevent overtopping of the ridges.

In these circumstances well-designed discharge points will be necessary at the furrow outlets. The size and spacing of the ridges should coincide with the crop's recommended spacing, furrow width and depth. Ridges and tied ridges may be constructed prior to, or after, planting. Maize is often planted on the flat, and the ridges constructed at the time of the first weeding about 30 days after planting, which saves labour. Clearly, the earlier ridges are constructed the more rainwater they will be able to detain, and the greater the probability of a good yield. The time when ridges are constructed is also a convenient time to simultaneously incorporate manures.

Advantages

The main advantage of contour ridges and tied ridges is the greater accumulation of rainwater within the furrows due to the retention of potential runoff. The concentration of water in the furrows encourages deeper percolation, but for this to be useful to the crop, the soil's AWC must be sufficiently high to retain the accumulated water within reach of the crop roots. Sandy soils with a low AWC may permit a large proportion of the rainwater to drain beyond the zone penetrated by the roots.

Disadvantages

The continual formation of ridges each year by hand or by mechanisation, combined with trampling along the furrows, may result in the formation of a compacted horizon at the base of the ridges, which can prevent roots from penetrating into deeper layers. This will counter the advantages provided by the ridges of increasing the supply of available water. The exposure of the soil surface leads to an accelerated loss of soil organic matter and surface crusting due to the effects of tillage, raindrop action and direct exposure to the sun, and very little macrofaunal activity. Consequently, the soils rapidly become degraded.

Another constraint is the time required to construct contour ridges, with even more time needed for tied ridges. The manual construction of contour ridges needs about 100 hours per hectare, and heavy textured soils will be even more demanding. To form ridges by hand or by animal traction in hardsetting soils will generally only be possible once the first rains have moistened the soil. The process of manually constructing contour ridges on sloping land, where the farmer faces uphill and pulls the soil into ridges with a hoe, causes soil to move downhill, so encouraging soil erosion.

Favouring Conditions

Contour ridges and tied ridges are most suited to areas suffering from water deficits where it is not feasible to provide a soil cover to enhance infiltration and reduce runoff through the use of crop residues, mulching materials or cover crops. The manual implementation of these structures will only be possible where sufficient labour is available and where farmers consider that the high labour requirement is justified by the value of the crop. These structures are particularly suitable for the production of tuber crops.

Impermeable and Permeable Contour Barriers

These structures include stone lines, walls, earth banks, fanya juu terraces, trash lines, live barriers and similar constructions. They have usually been installed to prevent small rills developing into gullies by limiting the area over which runoff collects, with or without sideways diversion into prepared waterways for safe disposal downslope.

The barriers, which they provide may, if well maintained, accumulate soil which has been eroded from upslope. In many situations, the chief benefit of laying out structures along the contour, at discrete intervals downslope, is their use as guidelines for the alignment of contour field operations in the cropping areas between them. The capture and soak-in of runoff along the upper sides of these structures may be considered as an added, rather than a primary, benefit.

The more closely spaced the banks, the more frequently runoff will be intercepted, but the more of the farmer's land will be taken out of production, unless some useful crop is planted along the earth bank. In semiarid areas structures can be designed for the purpose of water harvesting, which provides the extra water needed for adequate yields, if only from a relatively narrow strip.

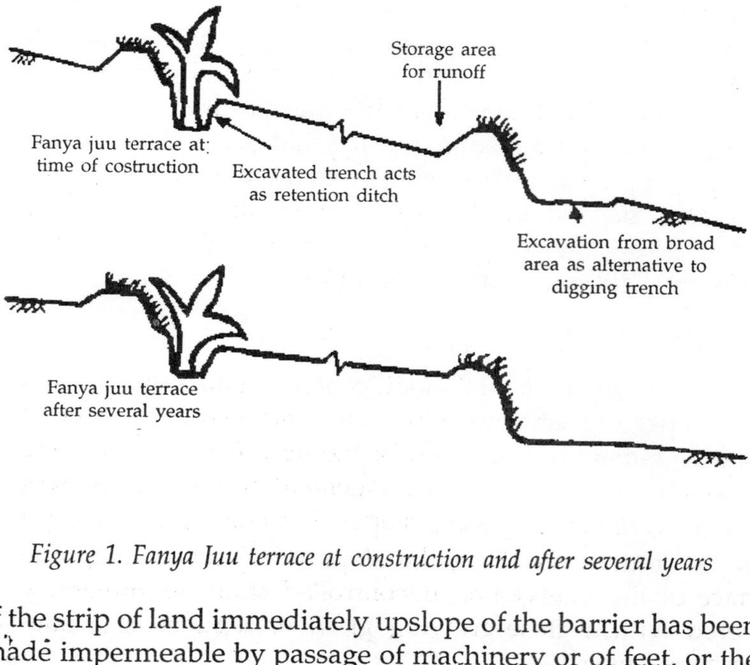

Figure 1. Fanya Juu terrace at construction and after several years

If the strip of land immediately upslope of the barrier has been made impermeable by passage of machinery or of feet, or the soil itself is relatively impermeable, temporary or more long-lasting local waterlogging can be induced. If more runoff accumulates than the structure can hold back on its upslope side, it will overtop and may break, with the resulting concentrated flow of accumulated water often causing more damage downslope than if the structure had not been there at all.

Favourable Conditions

In areas of moderate to high rainfall, such barriers may be appropriate where they complement water-absorptive conditions, good surface cover and/or ridge and furrows, with or without tied ridges. If they are laid out on the level contour they may have some small additional effect on increasing water in the soil.

Permeable Barriers

Permeable barriers, which may be accumulations of stalks, branches, crop residues, leaves (trash lines) without or with a line of one or more crops, forage grasses, shrubs or trees (live barriers) may impede but not stop runoff. The lower speed as runoff passes tortuously through the material provides an opportunity for infiltration. The live barrier may benefit from the additional soil moisture, but the additional transpiration through deep-rooted plants may minimise the volume, which could flow beyond the roots to groundwater. If the farmer receives no benefit from the live barriers, then the competition effects for light and moisture would be a disincentive.

Bench-type Terraces

These structures are a total modification of natural land slopes into a series of platforms which are almost level or slope at shallow gradient across or along the terrace. Controlling the gradient in this way allows management of water movement on what were formerly steep slopes. The cultivation platform may be continuous along the slope or discontinuous. The surface of the intervening uncontrolled slopes is preferably covered with a close-growing grass or legume as a soil-protecting cover.

Bench-type terraces are arduous and expensive to construct, requiring up to 700 person-days per hectare. Their capacity to receive and store rainwater depends on the depth, condition and quality of the soil into which they have been constructed. In semiarid areas they may be able to catch and detain all the rain that falls. In places with greater volume and frequency of rainfall, provision may have to be made for disposal of excess water down very steep waterways, and there is also an added danger of landslips if the benches become saturated.

Improving Subsoil Porosity and Permeability

Rainwater infiltration may be restricted in soils where the pore spaces rapidly become saturated with water because of the presence of dense subsoil horizons of low permeability. In these situations an initial deep tillage of the whole field with a tined implement, subsoiler or paraplow to break-up the dense horizon may improve subsoil permeability and so allow more rainwater to infiltrate. By improving subsoil permeability the rate of oxygen supply to the crop roots will also improve. However, the beneficial effects of deep tillage may only last 2 to 3 years.

Reducing Water Losses

The most effective solution to high evaporation losses of soil water is a cover of plant residues on the soil surface. Agronomic practices that increase shading of the soil surface, and physical structures that concentrate rainwater, encouraging percolation to deeper layers, also reduce evaporation losses. Wasteful transpiration losses may be the result of weeds or excessive crop transpiration in hot windy conditions, and can be reduced by appropriate weed control practices and windbreaks, respectively.

Minimising Evaporation

Surface residues reduce soil water losses through evaporation by acting as an insulating layer. This diminishes the temperature of the surface soil and eliminates the effect of wind. Heat from the sun is only slowly transmitted from the surface of the residues through the air trapped within the layer of residues to the soil surface. Consequently the soil surface remains cooler and the rate of evaporation of soil water is slowed down.

The thicker the layer of trapped air, the greater will be the insulating effect, and the quantity of residues required to reduce evaporation losses is considerably greater than the quantity needed to ensure that most rainfall infiltrates where it falls. For example, in Uganda, farmers traditionally apply between 8 and 40 t/ha of mulch to bananas, whereas 4-5 t/ha are probably sufficient to minimise runoff and allow most of the rainfall to infiltrate.

Banana yields respond very favourably to mulching, and applying 5-10 cm depth of maize stover and Paspalum sp. to bananas at Sendusu, Uganda, increased yields from 4.3 to 10.8 t/ha. Experiments in Uganda have shown that yields approximately doubled when 30-40 t/ha of mulch were applied compared with applications of 10-20 t/ha. This yield increase was mainly attributed to lowered evaporation losses. Protecting the soil surface from wind also slows down evaporation by reducing the rate at which water vapour is removed from the soil surface.

The use of residue covers for conserving soil moisture in the topsoil and increasing yields is particularly important in regions with limited rainfall and high evaporation rates. It is also important for shallow-rooted crops, e.g. bananas, tea, coffee, pineapple, and vegetables such as onion, lettuce, cabbage and carrots. Residue cover can also be very beneficial in reducing water losses by evaporation from soils with a shallow water table (less than 1 to 2 metres), from which there may be capillary rise of the subsurface water. Such soils are often used in horticultural production. However, the main disadvantage of using residue covers for reducing direct evaporation is the large quantities of residues required to significantly reduce evaporation. Often, the regions with high evaporation losses also suffer from a shortage of rainfall, which restricts the production of vegetative matter.

Frequently there are also other demands on residues, which take priority such as fodder, thatching and construction.

Reducing Transpiration

In hot windy weather, the rate of loss of water through plants by transpiration can be very high and can result in early depletion of limited soil moisture reserves. This in turn can lead to serious water stresses developing in plants—both crops and weeds—before their cycle of growth to maturity has been completed.

Weed Control

Loss of soil water through weed transpiration can seriously reduce the amount of water available to crops. Consequently, timely and effective weed control practices are essential. The presence of a thick layer of residues on the surface is a very effective way of controlling weeds. Where weed control measures are needed, the use of herbicides or appropriate crop rotations is often preferable from a conservationist perspective to mechanical weed control, unless it is practised with no soil disturbance.

Post-emergence herbicides leave weed residues on the soil surface as a protective cover whereas cultivation leaves soil exposed to the impact of raindrops and sun, accelerates drying of the surface soil and tends to disrupt and destroy soil porosity through smearing and compaction.

Windbreaks

When crops are exposed to strong winds in a dry environment the water that has been transpired by the crop is rapidly removed from the leaf surfaces into the atmosphere. This encourages a more rapid movement of water up through the crop and much greater absorption of water from the soil. Strong winds can therefore cause excessive crop transpiration rates and an unnecessary loss of soil water. Windbreaks will significantly reduce wind speed and so reduce crop transpiration rates and the unnecessary loss of soil water.

Windbreaks are usually established by planting single, double or triple rows of trees, but sugar cane or tall grass species may also be used. In areas where forests are being cleared for agricultural development, strips of the original forest may be left as natural windbreaks. Important considerations in the design of planted windbreaks are their composition, orientation, height, porosity and spacing. Windbreaks should be oriented at right angles to the direction of the prevailing winds during the growing season.

As a general rule, they should occupy no more than 5 percent of the cropped area. For small production units a single row of trees is usually most appropriate. Paths and roads should not cross windbreaks to avoid channelling of the wind through the openings at high velocities. The tree species selected should be adapted to the climate and soils of the area. The foliage should not be so dense that most of the wind is forced to pass over the top of the windbreak, as this will cause severe turbulence on the downwind side of the windbreak, which can seriously damage the crop.

The porosity of the windbreak vegetation should ideally be 40 percent so that part of the wind passes through the windbreak. This will give a 50 percent reduction in the velocity of the wind within a distance of ten times the height of the trees. When there is sparse protection in the lower part of the windbreak, as shown in Plate 54, it is advisable to allow regeneration of shrubs within the windbreak or plant tall grasses or sugar cane to ensure a more uniform protection from top to bottom.

Maintenance of the windbreaks is important to ensure that no holes appear, to regulate the porosity of the vegetation to wind and to avoid excessive shading and weed infestation of adjacent crops. Natural windbreaks are strips of forest left after deforestation. Since a much drier and windier microclimate develops in these strips of forest compared with that in the undisturbed forest, many trees in natural windbreaks often die, sometimes leaving holes through which the wind passes at increased velocity.

The important guideline for natural windbreaks, as for planted windbreaks, is that the porosity of the vegetation should be about 40 percent. In open forests in particular, natural windbreaks may need to be substantially wider than planted windbreaks to allow for the death of some trees. Alternatively, planting individual trees to fill gaps, or enriching the natural windbreak with one or two rows of additional trees may be necessary to produce a protective cover of 40 percent porosity.

Well-designed windbreaks will significantly reduce evapotranspiration rates of crops in windy conditions resulting in the conservation of soil water and less subsequent moisture stress when water is limiting. A 50 percent reduction in wind velocity (from 32 to 16 km/h) will reduce evapotranspiration rates by 33 percent. Windbreaks may provide additional benefits to crops by reducing mechanical damage and the loss of flowers, and by creating better conditions for insect pollination. They are also beneficial in reducing wind erosion, especially in fine-sandy and silty soils, and in diminishing air pollution problems.

Depending on the tree species selected, windbreaks may also provide fruit, nuts, fodder and timber, but the harvesting of these products must not result in pronounced gaps being formed within the windbreak. The main disadvantage for farmers with small plots is the loss of cropping area due to the windbreak and the risks of competition between the windbreak and the crop for water, nutrients and light leading to lower crop yields. This zone of competition may extend over a distance equal to 1.5 times the height of the windbreak. In areas where there are severe shortages of fodder, fuelwood and timber, windbreaks may need to be fenced to prevent indiscriminate grazing and harvesting. To ensure that wind cannot pass around the ends of individual windbreaks, the establishment of windbreaks should be planned on a community basis.

Conditions for Adoption of Windbreaks

Windbreaks will be favoured in areas subject to strong dry

winds during the growing season, and where windbreaks cause a net gain in soil water. Windbreaks are also likely to be favoured where they consist of species that provide additional benefits, such as fodder, fruit, nuts, fuelwood and timber that can be harvested without damaging the windbreak.

Shade

Shade can be provided by all manner of materials, whether artificial such as nets, cloths, plastic sheets and others, or plant-derived, such as cut branches, cut grass supported on nets, or living trees which provide high-level and wide-spreading shade. Shade is necessary in plant nurseries in hot regions to protect seedlings and other plants with shallow roots from rapid desiccation.

While shade may ameliorate the severity of hot dry conditions and limit undesirable losses of soil moisture, it can also be so dense as to limit solar energy reaching leaf surfaces and limit photosynthesis and growth rates. Where shade may be desirable, its density should be adjusted to provide an appropriate balance between losing water too fast, limiting sunlight intensity and avoiding scorching of leaves due to temporary dehydration and cell-damaging high temperatures. Using living shrubs and trees to provide long-term shade for tea and coffee can cause difficulties in maintaining the desired degree of shade above the crop over the long term.

Reducing Rainwater Drainage

In regions where much of the rainfall occurs as light showers, the concentration of rainwater as near as possible to the crop will cause more of the rainwater to infiltrate deeply, where it is less susceptible to evaporation. In order not to lose this water by drainage beyond the crop's rooting zone and where there is no rooting restriction some solutions can be adapted, such as increasing the capacity of soils to retain water within the rooting zone, early planting to accelerate root development or changing to deeper-rooting crops.

The addition of large quantities of organic manure will increase the available water capacity (AWC) of soils and in theory this is a useful practice for reducing deep drainage losses. However, even in temperate climates the quantities of organic materials required to markedly increase AWC are very high, applications must be continued over many years and usually affect only the plough-layer depth.

In tropical zones, where organic matter decomposition rates are much higher, the influence of organic manures on AWC is likely to be even less. Nevertheless, this practice may be feasible for small-scale farmers growing high-value crops where large quantities of organic manures and labour are readily available. In low rainfall areas, it is frequently difficult to know when the rains have truly started, as initial rains are often followed by a dry period.

Many farmers wait until the topsoil has been moistened to a depth of about 15-20 cm before planting, so that even if there is a subsequent short dry period there is sufficient water within the soil. However, this results in a delay in planting and for every day's delay yields will decrease, largely due to the loss of rainwater by drainage and evaporation, together with the loss of some released nutrients. To overcome this problem and to allow crops to develop deeper rooting systems earlier on so that more of the rainfall can be utilized during the initial stages of the season, some farmers "dry plant" when soils are dry prior to the onset of the rains.

To avoid premature germination before sufficient rain has fallen, the seeds are usually placed deeper than normal. Dry planting also has the advantage of spreading labour over a longer period. Crops may also benefit from this practice by being able to utilise the nitrogen released at the start of the rains from the decomposition of soil organic matter, which reduces leaching and pollution of groundwater. However, there are a number of problems associated with dry planting, notably that some soils, and in particular hardsetting soils, are difficult if not impossible to till when dry.

If seeds are not planted sufficiently deeply, they may germinate at the first rains and then die during a subsequent dry period. Applying fertiliser to speed up crop canopy development and increase the shading of the soil surface will decrease the soil water lost by evaporation so that more is available to the crop. Planting crops equidistantly so that the soil surface becomes shaded more quickly would also be expected to reduce the proportion of soil water lost by evaporation. However, the effects of these agronomic practices on reducing evaporation losses will be much less than applying surface residues.

On permeable sandy soils that retain small quantities of available water for crop use, it is preferable to introduce deep-rooting crops that can utilise soil water at depth that would not be available to shallow-rooting crops. Examples of deep-rooting crops are almond, barley, cassava, citrus, cotton, grape, groundnut, olive, pearl millet, pigeon pea, safflower, sisal, sorghum, sunflower, sweet potato and wheat.

Improving Soils with Restricted Rooting

The type of solution to be applied will depend on the cause of root restriction. The most frequent cause is physical root restriction due to a lack of pores that are large enough to be readily penetrated by roots or which can be sufficiently widened by the growing roots. This condition occurs in dense layers, such as plough pans formed by tillage, but also in naturally occurring dense layers as found in hardsetting soils. Root restriction may be overcome, at least temporarily, by biological or mechanical means.

In addition to eradicating the causes of root restriction it is also important to take steps to avoid future recurrence of the problem by, for example, introducing conservation agriculture where dense layers have been formed by tillage. Less common causes of restricted rooting are chemical restrictions due to the presence of toxic concentrations of aluminium or manganese, high salinity or severe nutrient deficiencies, especially of phosphorus. A lack of oxygen due to a fluctuating water table may also restrict root development.

While the water table is high, root development for most crops will be restricted to the soil immediately above the upper level of the water table but the crop will not suffer from a lack of moisture. If the water table then falls relatively quickly to a substantially lower level, for example at flowering, when the crop has still to reach physiological maturity but the roots have ceased growing, the roots may be left stranded in the dry soil without access to the moisture in deeper layers.

The causes of restricted rooting given above can, where appropriate, be overcome by the application of lime, or lime and the more mobile gypsum, to eradicate aluminium and manganese toxicities; leaching to reduce salinity hazards; fertilisers to rectify nutrient deficiencies; or drainage to remedy the lack of oxygen from a fluctuating water table. The principal biological method of restoring the porosity of root-restricting layers is to place the land in fallow and utilise the roots of natural vegetation or planted cover crops to act as biological subsoilers penetrating the dense root-restricting horizons.

The stability of root channels created by plant roots will be greater than that of channels formed by mechanical methods because of the release of organic substances from the roots that stabilise the channel surfaces. Once the roots have died and shrunk, these pores will be sufficiently large and stable to enable the roots of subsequent crops to penetrate. Land may be left in fallow for 2-3 years for natural bush or forest vegetation to regenerate. Alternatively, planting selected species that are effective in regenerating soil structure can enrich the natural fallow.

A cover crop may be sown to serve as a planted fallow. Promising cover crop species that have been shown to have potential as biological subsoilers are the grasses Bahia grass (Paspalum notatum), Festuca elatior, Guinea grass (Panicum maximum), and alfalfa (Medicago sativa), pigeon pea (Cajanus cajan) and cowpea. Radish and the nitrogen-fixing shrubs Tephrosia vogelii, Sesbania sesban and Gliricidia sepium have also been identified as potentially useful. Some weeds with pronounced tap-roots, such as Amaranthus sp., may possibly also have potential to act as biological subsoilers, as Mennonite farmers in eastern Bolivia have observed much higher crop yields on compacted soils after high infestations with Amaranthus.

Biological methods are generally much cheaper to implement and their benefits are longer-lasting than mechanical methods. An important advantage of vegetative fallows is that they greatly improve the physical, chemical and biological fertility of the soil due to the large quantities of organic matter produced and added to the soil. Tree fallows can be beneficial in supplying fuelwood, construction materials and other products, provided the harvesting of these materials does not reduce the beneficial effects of the fallow on soil chemical fertility.

The main disadvantage is the 2 to 3 years required for natural fallows when the land is taken out of production while the recuperation takes place. A disadvantage of tree fallows is the difficulty of returning to annual cropping after the fallow period because of the problem of extracting the tree roots and the longer the fallow period the more difficult the problem. However, the extraction of the roots of Sesbania after 2 years of fallow has not been a problem in Zambia.

It is also necessary to protect the vegetation from grazing, burning and harvesting during the 2-3 year fallow period, which may involve additional costs for fencing. Planted fallows of cover crops with tap-roots may be difficult because of the lack of available seeds and their cost, since a high plant population is necessary to ensure an adequate density of tap-root penetration of the root-restricting layer. For very dense root-restricting layers, even Cajanus cajan may have only a limited effect.

Favouring Conditions for Biological Methods

The use of natural biological methods will be favoured by farmers who have sufficient land. They can take some of it out of production and place it into fallow while the slow process of natural regeneration of soil porosity takes place. The use of cover crop fallows is often a rapid process which enables land to be more quickly returned to production. Natural fallows in which there is a regeneration of tree vegetation are more likely to be adopted by farmers who wish to change the land use of the recuperated area to forest or perennial tree crops.

Mechanical Methods to Root Restriction

The aim of mechanical methods is to break-up the compacted or naturally dense root-restricting layer in order to create larger pores through which crop roots can penetrate. This is accomplished by the implement slightly lifting and breaking the compacted or dense layer. The operation may be carried out over the whole of the field, or merely along the rows where the crop is to be planted.

The latter, known as in-row subsoiling, is much quicker and requires less draught power, but the crop must be sown with precision directly over the loosened rows. The most appropriate method will depend on the depth to the root-restricting layer, its thickness and hardness, and the source of power available.

Shallow Root-restricting Layers

Shallow root-restricting layers such as hoe pans are typically produced at 5 to 8 cm depth, and the easiest means of breaking them up are with ox-drawn rippers or tractor-mounted chisel ploughs. Most farmers relying on manual tillage will probably have to use hand tools to break the hoe pans by methods such as double digging, which are very labour-intensive. To break-up compacted layers in the dry season when the soil is very hard may require robust tools different from those the farmer normally uses for tillage, such as pickaxe, mattock, three-tined hoe or a long crowbar.

In central and western Kenya, small resource-poor farmers intensified their production, both in yield and diversity, by using double-dug (to 50 cm depth) composted beds on small areas, generally near to their houses. Positive results were achieved from the concentration of organic materials onto the beds, which received focused attention, plus improved rainwater capture. Improved conditions in the root zone, including excellent moisture-holding capacity, enabled a range of vegetables (and field crops) to be grown well into the dry season, and these were less affected by drought than those grown in unimproved plots

While the total area of land managed in this way is often only a small proportion of the total cropped land, overall output from the beds rose sharply due to higher yields and diversification of crops. The system provides many benefits which were recorded during a survey of farm families' comments. Deeper root-restricting layers such as plough pans are formed at the lower limit to which the soil is tilled, and usually occur within the upper 10 to 25 cm of the soil profile.

Plough pans formed by ox-drawn implements can usually be broken up using two passes of an ox-drawn ripper, whereas those formed by tractor-drawn or -mounted implements usually require a tractor-mounted subsoiler or paraplow. Paraplows are similar to subsoilers except that the shanks are slanted sideways to the direction of travel, which enables soil to flow over the shanks. They are preferable to subsoilers as they bring fewer subsoil clods to the surface, require less draught power and cause less incorporation of surface residues that should ideally be left on the surface.

Disc ploughs are less suitable because they invert the soil, incorporate most of the crop and weed residues and bring subsoil clods to the surface, resulting in the need for additional tillage. If the root-restricting layer is to be disintegrated over the whole field, then as a rule of thumb the subsoiler or paraplow should penetrate to 1.5 times the depth to the lower limit of the root-restricting layer, and the spacing of the shanks should not be greater than this value. For example, if the root-restricting layer occurs at 10-24 cm depth, the shanks of the subsoiler or paraplow should penetrate to 36 cm and the spacing between the shanks should be no more than 36 cm.

If the shanks are more widely spaced, there is a likelihood that the root-restricting layer will not be fully disrupted in the region midway between where the shanks passed. To avoid compaction from the wheels of the tractor, shanks should be positioned immediately behind the tractor's wheels. For in-row subsoiling, the shanks need only penetrate to the lower limit of the root-restricting layer, and the shank spacing should coincide with the planned row spacing of the crop. Subsoiling should be carried out perpendicular to the normal direction of tillage and the soil should be dry to the depth of subsoiling to obtain good shattering. If the soil is moist or wet, there will be no shattering, merely the formation of channels gouged out where the subsoiler's points have passed.

Mechanical Disruption of Very Deep Root-restricting Layers

Subsoil compaction at 40 cm depth and greater is caused by the passage of very heavy equipment with high axle loads of at least 6 tonnes, such as combine harvesters and lorries laden with grain. At this depth the use of conventional subsoilers to loosen deep compacted layers is difficult and expensive because of the very high traction power needed. Vibratory and rocking subsoilers, in which the subsoiler points vibrate or rock using the tractor's power takeoff can work to 80 cm depth, but require 75-100 HP.

New implements have been developed employing elliptically moving blades or rotary hoes, which utilise a break-off-loosening mechanism to disintegrate compacted layers. They can be used to depths of 60 to 120 cm and at higher soil moisture contents than conventional subsoilers, but are very expensive and require high traction power. The shattering and lifting of root-restricting layers by mechanical means creates larger pore spaces through which roots can penetrate, enabling them to reach and take advantage of soil moisture and nutrients stored in deeper layers.

Consequently, crops are able to make more efficient use of the rainfall. The main effect of subsoiling is usually that of promoting deeper root growth, but if the root-restricting layers are so dense that rainwater movement is also limited, subsoiling may also facilitate the percolation of rainwater into deeper layers. The development of improved rooting frequently increases crop and pasture yields. In Babati District, Tanzania, breaking up hardpans by subsoiling has almost tripled maize yields and quadrupled maize dry matter production. Increased yields from subsoiling are most likely in areas where yields are limited by rainfall, and the drier the season the greater the probable response to subsoiling.

In-row subsoiling, especially when it is combined with planting in a single operation, is particularly beneficial for hardsetting soils that rapidly form root-restricting layers on drying after being saturated with rain. This technique is most likely to be successful when associated with precision planting and controlled traffic, in which the passage of all machinery wheels is restricted to permanent tracks. The benefits of subsoiling are likely to be greatest when immediately followed by the establishment of a dense cover crop with a strong rooting system that helps stabilise the new pore spaces created.

The cover crop should then be followed by a system of conservation agriculture in which the absence of tillage reduces the recurrence of further compaction. The principal disadvantage of mechanically breaking up root-restricting soil layers is the high power requirement, whether it is manual, animal or mechanical. Since most farmers do not have access to more than that which they use for land preparation, the process is inevitably slow. Some soils become so extremely hard during the dry season, that the farmer's normal draught power is incapable of penetrating the soil in order to break-up the root-restricting layer.

It is then necessary to wait for the beginning of the rains to moisten and soften the soil before it becomes possible to break-up the compacted layer, but this may coincide with the critical time of land preparation and planting. This problem can apply equally to farmers using animal traction or tractors and to those using hand tools as their source of power. Subsoiling operations are ineffective when the dense or compacted layers are wet or very moist as no shattering effect takes place.

Farmers often lack the necessary implements, whether they are pickaxes for farmers relying on manual power, rippers for animal traction farmers, or subsoilers or paraplows for mechanised farmers. The use of normal land preparation implements will generally be less satisfactory. For example, disc ploughs can be used to break-up plough pans, but they invert the soil bringing large clods of subsoil to the surface and form an uneven surface that needs additional tillage to create a seed bed.

Disc ploughs also incorporate the residues of crops and weeds, when ideally they should be left as a protective layer on the soil surface. Repeated use of disc equipment, especially heavy-duty disc harrows, can produce an almost impermeable compacted pan in only a few seasons. These pans have been the cause of increasingly severe runoff and erosion from millions of hectares in Brazil, before the use of disc equipment was abandoned in favour of minimum tillage with tines, and subsequently by no-till systems.

When bulky crop residues are left on the surface, especially the stiff residues of maize, sorghum and cotton, the performance of subsoilers and paraplows is considerably impaired unless they are fitted with front cutting discs. If subsoiling is followed by conventional tillage, the beneficial effects are only likely to persist for 2 or possibly 3 years and so the subsoiling has to be regularly repeated. The speed with which the root-restricting layer reform will depend on the number of tillage and other field operations, the moisture content of the soil at the time of these operations and the susceptibility of the soil to compaction.

Fine-sandy and silty soils and those with impeded drainage are most susceptible to compaction. To improve the physical conditions of hardsetting soils requires the incorporation of large quantities of organic material into the dense layers and the regeneration process is likely to be slow. For hardsetting soils, in-row subsoiling may be necessary each year. These disadvantages can be overcome by adopting reduced tillage, or preferably zero tillage as in conservation agriculture, or by controlled traffic in which all machinery follows the same tracks year after year, leaving the cropped strips untouched.

Thorough loosening of soils by subsoiling may render them more susceptible to compaction if they are subsequently subjected to high pressures, as from excessive tillage or the passage of very heavy machinery. The recompaction may be worse than the original state of compaction. Subsoiling heavy textured soils, such as vertisols, can greatly increase the quantity of rainwater that reaches the subsoil, resulting in a marked reduction in the soil's bearing strength, i.e. its capacity to support heavy machinery.

It should be noted that subsoiling to any given depth produces a high proportion of very large soil pores and fissures, a situation favouring better penetration of roots and of rainwater. It will not however produce any significant increase in the range of smaller soil pores, which make up the water-retention capacity of the soil.

Favouring Conditions

The adoption of mechanical methods to overcome physical root restriction will be favoured where yields are frequently limited by low rainfall. Under such conditions it becomes important that as much of the rainfall as possible is stored within the soil profile, and that the crop's roots have access to all of the stored soil moisture. Mechanical methods will be favoured where farmers have access to tractors and subsoilers or paraplows, and where land cannot be taken out of production and put down to fallow for 2 to 3 years.

Chemical Solutions

Root development is sometimes restricted by unfavourable soil chemical conditions, such as severe nutrient deficiencies, aluminium or manganese toxicity and salinity. The nutrient which most commonly restricts root development is phosphorus and the application of P fertilizers to phosphorus-deficient soils frequently encourages deeper rooting, enabling the crop to access more soil moisture and so increase productivity. The incorporation of lime without or with gypsum will reduce toxic concentrations of aluminium and/

or manganese to non-toxic levels and so encourage deeper rooting.

The greater solubility of gypsum compared with lime makes the former more suited to soils with aluminium or manganese toxicity problems in the subsoil, whereas the slowly soluble lime is most effective in topsoils. When high salt concentrations inhibit root development in irrigated soils, excess quantities of water should be applied sufficient to leach the salts out of the crop's rooting zone.

Usefulness of Low and Erratic Rainfall

Several approaches may be used to diminish the impact of low and erratic rainfall, viz. match land use to soil characteristics; use drought-resisting and drought-escaping crops; increase the efficiency with which crops utilise rainwater; concentrate rainfall by water harvesting; divert river water; intercept floodwater; and apply supplementary irrigation.

Match Land Use

Matching land use to the most suitable soil types within a farm may increase the efficiency with which the available soil water in the different soil types is utilised for crop production. Crop water requirements vary, as do the capacities of soils to retain and supply water to crops.

Table 1. Length of growing period for different soil available water capacities in bimodal rainfall areas of semiarid India

	Length of growing period (weeks)		
Rainfall probability	Low AWC 50 mm (Shallow alfisol)	Medium AWC 150 mm (Medium vertisol)	High AWC 300 mm (Deep vertisol)
Mean	18	21	26
75%	15	19	23
25%	20	24	30

Moreover, the variations in available water capacities (AWC) of soils often occur over short distances. Soils with high AWC will be expected to suffer less water loss from deep drainage and possibly from runoff. Consequently, greater quantities of rainwater will remain in the soil and so the potential crop-growing season will be longer assuming an adequate amount, distribution and infiltration of rainfall.

The longer the expected duration of dry periods and the more sensitive the crop to drought, the more important it will be to use soils of high AWC. For soils to be considered suitable for maize in semiarid areas of Arusha, Tanzania, they must be of sufficient depth and AWC for the maize to be able to tolerate dry periods of up to four weeks. Farmers can take advantage of variations in AWC by locating moisture-sensitive crops and crops with longer growing periods on soils of high AWC and crops tolerant to drought and early-maturing crops on soils of low AWC.

This approach is applicable at farm level and also at field level, especially for farmers with very smallholdings where differences in soil AWC between small areas within a field can still permit diversification. Some localised areas may occur within a field where runoff accumulates and provided the soil's AWC is adequate to retain the moisture, the soil will be suitable for more water-demanding crops.

Seasonally waterlogged low-lying, grassy areas, known as dambos, are commonly found at the head of watercourses in southern and central Africa. Their high soil water content makes them highly suitable for crop production, even in semiarid areas, because they are relatively unaffected by mid-season droughts. Even in dry years, yields up to 2.5 t/ha of maize can be obtained. Traditionally, dambos were used for rice, maize and vegetable production, dry season grazing and sources of domestic water.

In Zimbabwe the cultivation of dambos was banned because of concern about environmental degradation, but recent research has shown that with environmental safeguards, present levels of yield could be increased threefold. Matching crops with weak root systems, such as beans, to soils lacking root-impeding layers, would be expected to increase crop water use efficiency. Beans are more suited to freshly tilled soils, or to mature no-tilled soils where large numbers of channels suitable for root penetration have been created through the decomposition of old roots and soil faunal activities.

Allocating land use to suitable soil types may enable production to be intensified, leading to benefits in addition to that of higher water use efficiency. Thus, intensifying subsistence food crop production may liberate land for producing cash crops. Alternatively, it may allow land previously used for inappropriate, extensive and degrading forms of land use, to revert to natural vegetation, thereby reducing land degradation.

Drought-resistant and Drought-escaping Crops and Varieties

Some crops can tolerate drought because they are able to resist a shortage of water, i.e. they are said to be drought-resistant. This is because either:

— they can stop growing when water is unavailable by becoming dormant. When rain occurs they resume growing and developing as though nothing had happened; or

— they have deep rooting systems, such as pigeon pea, and can absorb water from deep within the soil. This is important where occasional rainstorms wet the soil to great depth followed by long dry periods.

Pineapples and sisal resist the effects of drought due to their thick leaves that slow down water loss by transpiration. These crops, as well as sorghum, pearl millet, pigeon pea, cassava, groundnut and cowpea, are drought-resistant and suited to climates with a defined mid-season drought. Drought-escaping crops are those that can tolerate droughts because they have short growing periods and mature quickly before all the soil water has been used up.

Early-maturing cultivars have been successfully bred in Kenya for dry areas, such as Katumani Composite maize and "Mwezi moja" beans. Cowpeas mature early and are both drought-escaping and drought-resistant. A drawback of drought-escaping crops is that their short growing season restricts yields compared with long-season cultivars, although under dry conditions they will outyield the long-season cultivars. For example, improved pearl millet varieties in Tanzania, which mature two weeks earlier than farmers' local varieties, have yielded 43 percent more (2.31 t/ha) than local varieties (1.62 t/ha).

Applying fertilizers to counteract nutrient deficiencies can speed up crop maturity and so enable them to escape droughts more easily. Drought-escaping crops are more suited to short rainy seasons, or to soils which can only store a limited quantity of water. It is therefore important to select drought-escaping crops and varieties whose maturation period matches the expected length of growing season. If possible very determinate varieties should be avoided so that the risks of the whole crop being adversely affected by dry periods is reduced.

Unfortunately, farmers often do not have access to varieties that match the expected length of growing season, and the length of growing season may vary widely from year to year. It must be borne in mind that the choice of crops and varieties depends not only on their ability to resist or escape droughts, but also on their susceptibility to pests and diseases, labour requirements, availability of seed, ease of grain processing, fuel requirements for cooking, and palatability.

Increase Crop Water Use Efficiency

Crop water use efficiency refers to the amount of dry matter produced for each millimetre of water that is transpired by the crop or evaporated by the soil, i.e. for each millimetre of evapotranspiration. Clearly, in dry areas the more efficient use the crop can make of the rainfall that infiltrates (referred to as the effective rainfall), the higher will be the yield. The following management practices influence crop water use efficiency.

Selecting Water-efficient Crops

A group of crops referred to as C4 crops, which include maize, sugar cane, sorghum and pearl millet, are physiologically much more efficient at producing dry matter for each millimetre of transpired water than other crops, referred to as C3 crops. But this distinction is most important in situations where rainfall is adequate. For areas where water deficits are common, the use of drought-resistant and drought-escaping crops is much more important.

Adjusting Plant Population to Expected Rainfall

A high plant population will use large amounts of water for transpiration during early growth provided sufficient water is available in the soil. Because of rapid shading of the soil by the crop foliage, less water will be lost by direct evaporation, ensuring a higher water use efficiency compared with low plant populations. High plant populations, and especially those with a more square planting arrangement, also increase water use efficiency through the quicker development of cover and therefore less weed growth.

Although evaporation losses are greater for low plant populations, soil texture and the frequency of rainfall events also influence the amount of water lost. Sandy soils in areas where rainfall occurs in few heavy storms will suffer less evaporation than medium or fine textured soils in areas with frequent rainfall events.

Where rainfall is erratic the situation is complicated, and becomes more than just a matter of water use efficiency. Farmers then face the dilemma of whether to sow at a low density to ensure some yield in bad years but underperforming in good years, or to use a high population to maximise yields in good rainfall years but to harvest very little, if anything, in bad years. If farmers have sufficient land they can opt for both options, i.e. an area with low population and another area with high population, but many small-scale farmers possess insufficient land for this to be feasible.

Response farming is an approach for matching crop management to estimated seasonal rainfall in variable rainfall zones. Plant populations and N fertilizer applications are adjusted after the crop has been established on the basis of information about the expected rainfall. Initially, the crop is sown at a high population assuming a good rainfall season, and with a low application of N fertilizer.

The expected potential of the season (good, fair, poor) is determined on the basis of the anticipated amount of rainfall during the first 30-50 days, derived from as many years' records as are available. Decisions are then made according to the amount of rainfall early in the season on whether or not to thin or to apply additional N fertiliser. So far, this practice has not been adopted by farmers because of the great variations in seasonal rainfall over short distances, because farmers usually intercrop, and because of the initial wastage of water that occurs if crops are thinned after 30 or more days.

References

Blatter, Joachim and Ingram, Helen, eds., *Reflections on Water: New Approaches to Transboundary Conflicts and Cooperation,* Cambridge, Mass.: MIT Press, 2001.

Kliot, Nurit, Shmueli, Deborah, and Shamir, Uri., *Institutional Frameworks for the Management of Transboundary Water Resources,* Haifa, Israel: Water Research Institute (Two volumes), 1997.

Milich, Lenard and Varady, Robert G., "Openness, Sustainablility, and Public Participation: New Designs for Transboundary River Basin Institutions," *The Journal of Environment and Development* 8(3): 258–306, 1999.

Savenije, Huber H. G. and van der Zaag, Pieter., "Conceptual Framework for the Management of Shared River Basins; with Special Reference to the SADC and EU," *Water Policy* 2(1-2):1–37, 2000.

Wolf, Aaron T., "Criteria for Equitable Allocations: The Heart of International Water Conflict," *Natural Resources Forum* 23:3–30, 1999.

8

Improving Soil Fertility

The most striking property of Amazonian Dark Earths is their high fertility. This has been described in several publications since the early expeditions in the 1860s and 70s by Cornell professor Charles F. Hartt and his student Herbert H. Smith and analytical data have been documented by Sombroek, Smith, Kern and Kämpf and others. However, these soils do not have high availability of all nutrients necessary for optimum plant productivity. In addition, total contents may be high but they are not available to plants. The latter is, for example, true for nitrogen (N). The N contents of ADE are typically much greater than those of adjacent soils, however the availability of N may be only slightly higher or even lower, as shown in the example from Klinge.

For two ADE from the lower Rio Negro and Belterra, available N increased more than total N in comparison to adjacent soils, but N mineralisation in relation to total soil N was lower in the Dark Earths. Additional evidence is provided by a greenhouse experiment, in which foliar N contents and total N uptake by rice and cowpea were 18-58 % lower when grown on ADE than on a Ferralsol. Total and available phosphorus (P) and calcium (Ca) contents usually increased the most in ADE compared to other nutrients. Available P increased more than total P, indicating that a larger part of the added P in the Dark Earths is in plant-available form compared to adjacent soils.

Available and total Ca and magnesium (Mg) contents do not show a consistent trend in this respect. Available manganese (Mn) contents increased to the same extent as total Mn. Neither available nor total potassium (K), iron (Fe), and zinc (Zn) increased compared to adjacent soils. Both total sulfur (S) and copper (Cu) contents increased however, by more than 100%. The presented data reflect general trends for the different nutrients. Comparisons may not be clear-cut in all situations because it is difficult to determine where ADE ends and where the adjacent and unenriched soil begins.

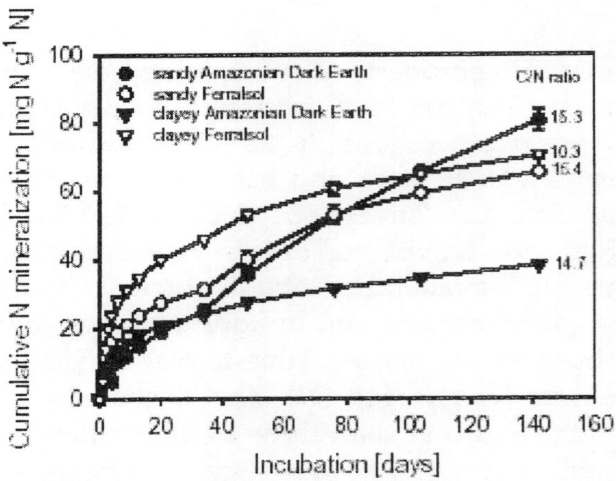

Figure 1: Cumulative N mineralization (NO_3^-, NH_4^+, organic N) of sandy (88% sand) and clayey (10% sand) Amazonian Dark Earths in comparison to adjacent Ferralsols; determined in mini-lysimeters of soil mixed with quartz sand (1:1 w/w)

The largest proportion of N is usually present in soil in organic forms, and this is also the case for ADE, while more inorganic than organic P is usually present. The examples shown from cultivated fields indicate that the proportion of inorganic N and P increases in ADE compared to adjacent Ferralsols. The uncultivated fields showed lower proportions of inorganic than organic P in ADE than Ferralsols. No information is currently available for S.

These data indicate that Amazonian Dark Earths do not necessarily have high availability of all nutrients important for plant growth. While total and available contents of some nutrients, such as K, Mg, or Fe, may not be higher at all, other nutrients have higher total contents, such as N, but those contents are not necessarily readily available. The mineral matrix of Amazonian Dark Earths is strongly weathered, often showing high contents of quartz and kaolinite contents, and therefore possesses few weatherable minerals which could potentially release nutrients. Thus, nutrient availability in ADE relies more on nutrient release from exchange sites and soil organic matter through biological processes than on weathering of parent material.

Nutrient Release from SOM

Nitrogen, P, and S are contained to a large extent in soil organic matter and their release from SOM determines their availability to plants. This release depends on the so-called "quality" of SOM. For example, the C/N ratio of the clayey Ferralsol (10.3) was lower than the C/N ratio of the clayey ADE (14.7), and consequently cumulative N mineralization was higher in the Ferralsol than the ADE. Nitrogen, P, and S are released more slowly or even immobilised by microorganisms, if the C/N, C/P, or C/S ratios are above 20, 200, and 200, respectively.

The C/N ratios of Amazonian Dark Earths reach from 12-40, and are frequently higher than those of other upland soils in the Amazon Basin. However, important differences exist between nutrient release from soil with similar ratios, since the chemical forms of C and its physical protection in aggregates and organo-mineral complexes are known to affect SOM mineralisation and the concurrent release of nutrients. A higher proportion of N was stabilised in the silt fraction of ADE than that of Ferralsols, indicative of stable organic nutrients. Therefore, net N mineralisation in ADE may not only be low due to high C/N ratios but also due to physical protection of organic N.

Also the proportions of stable organic P and of those P fractions with slow turnover, as obtained from sequential P fractionation, were higher in ADE than in the cultivated

Ferralsols. At the same time, P turnover was higher in ADE than in the adjacent Ferralsols, as seen from a high proportion of P in microbial biomass and higher respiration. Thus, N is more likely to be immobilised and less available in Amazonian Dark Earths than P, which shows high availability for plant uptake.

Another consideration is the high amount of black carbon present in ADE and its effects on N, P, and S release. In the absence of direct experimental evidence, we can hypothesize for N that:

(i) black C particles have high C/N ratios and do not release any N, but do not influence the N release of non-black C. In this case, the N release from the organic matter that is not associated with black C would need to be determined separately;

(ii) black C is finely distributed and has a direct influence on the N release from non-black C.

Available data indicate that black C indeed influences total N release either by microbial immobilisation due to the high C/N ratios of the black C or by catalytic effects of the high surface area of the black C.

Adsorption of Nutrients

The adsorption of nutrients by soils affects nutrient retention against leaching and plant-availability. Amazonian Dark Earths often show higher soil pH, coarser soil texture, higher SOM contents and altered SOM quality than adjacent soils, which all change adsorption properties. Higher pH usually increases the effective cation exchange capacity (CEC determined at the pH of the soil) especially of highly weathered soils with a high proportion of variable charge. Higher quartz contents and lower kaolinite contents of ADE than adjacent Ferralsols decrease the CEC of the mineral matrix.

This may not play an important role, however, since the CEC of the clay minerals is very low in acid upland soils in Amazônia and the total CEC is mainly controlled by SOM. At the same time, lower clay contents may decrease fixation and occludation of P. This has not yet been directly shown, but can be expected. The usually higher SOM contents in ADE increase the potential CEC (determined at pH 7), and the higher pH increases the effective CEC and base saturation.

In contrast to soils from temperate climates, which usually exhibit a strong correlation between clay content and CEC, only 33% of the CEC increases in Amazonian Dark Earths can be described by their clay contents ($y=0.698x+13$; $r^2=0.33$; $P<0.001$; $N=55$ from topsoils of independent sites). However, this relationship is noteworthy since Sombroek showed that the clay does not contribute at all to CEC in the highly weathered upland soils of central Amazônia. The reason for this trend towards higher CEC with increasing clay contents in Amazonian Dark Earths may lie in the fact that higher clay contents increase aggregate formation and C stabilisation.

Additionally, SOM in Amazonian Dark Earths contributes proportionally more to CEC than in Ferralsols with comparable organic matter contents. This result indicates that SOM in ADE is more effective in increasing CEC than SOM in Ferralsols. An explanation may be found in the fact that ADE contain large amounts of partly oxidised pyrogenic carbon (black carbon), which may have more exchange sites than non-pyrogenic decomposed plant material due to oxidation of the aromatic carbon and formation of carboxylic groups.

Additionally, the pyrogenic carbon is more stable than the non-pyrogenic carbon. As a consequence, more exchange sites are formed and maintained before the organic material is mineralised to carbon dioxide. Sombroek provided data showing a significant relationship between total P and potential CEC, which were interpreted by a formation of phosphate complexes in organic matter near the topsoil and with phosphate bonding to mineral surfaces in the subsoil. No direct evidence can be found to date to support a direct complexation of phosphate and organic matter.

An alternative explanation is that organic matter is usually strongly bonded to Fe oxides, which in turn fix inorganic P creating the observed correlation between P and CEC. High organic matter usually affects the adsorption of P and micronutrients. However, very little is known about the specific effects in ADE, and P adsorption isotherms have not been determined. Organic acids were shown to decrease P adsorption and thereby increase P availability in a Brazilian Ferralsol. This effect may contribute to the proportionally high P availability in relation to total P observed in the Amazonian Dark Earths. Falcão et al. observed high P adsorption to charcoal collected from ADE which may explain their high P retention and concentration. It is interesting to note that plant-available P was not related to the amount of organic C, whereas 48% of the variability of the extractable Ca could be explained by the C contents of Amazonian Dark Earths.

Additionally, Ca and P were not significantly related ($P>0.1$; $N=60$; topsoils of independent soil profiles). None of these relationships were affected by soil texture. Therefore, ADE with high C contents generally contain more available Ca, but this is not necessarily true for P. Similar significant correlations between organic C and Ca were not found for clayey Ferralsols from central Amazônia (A. Moreira, unpubl. data; 15-80 g C kg^{-1}; available Ca 1-21 mmolc kg^{-1}; $N=505$), or from Ferralsols and Acrisols of the USDA data base. This result has implications for the classification of Amazonian Dark Earths as well as for their fertility management.

Further implications may be provided with respect to the discussion of the origin of ADE and the sources for their extraordinarily high C, Ca, and P contents. Two explanations emerge from the observed relationships: either (i) at least two different materials were added to soil, one with both high C and Ca contents, and one with high P contents; or (ii) the stabilisation of C, Ca and P in soil organic matter followed different patterns. We will not be able to elaborate on the first point in this contribution, but we will discuss the second explanation. It is known that the Dark Earths were enriched with organic matter and nutrients at least 500 years ago.

Therefore, sufficient time passed to allow significant changes in soil chemistry. Two processes may explain the close relationship between Ca and C, (i) a stabilisation of C by Ca, and (ii) a stabilisation of Ca by C. Soil organic matter extractability was shown to be lower after additions of Ca-hydroxide to a humic Ranker from Spain or $CaCO_3$ and $CaSO_4$ to an Alisol from Australia, but mechanisms and importance of the stabilisation remain uncertain.

Additionally, Ca and soil organic matter were frequently shown to increase soil flocculation in contrast to, for example, Mg to improve aggregate stability and may have enhanced physical protection of SOM. However, the opposite is often observed in acid upland soils in the humid tropics, which have a low point of zero net charge and exchange sites dominated by Al. Schuffelen and Middelburg pointed out that increasing the pH by additions of lime decreases aggregate stability due to a displacement of the trivalent Al by the divalent Ca. These results suggest that Ca is unlikely to increase C stabilisation to a large extent.

On the other hand, there may not only be an effect of Ca on C contents, but C may also affect Ca stabilisation. Organic matter increases the CEC which improves Ca retention. Since Amazonian Dark Earths contain large amounts of Ca, this may explain the observed relationship between Ca and C. However, Ca contents are several orders of magnitude greater than in adjacent soils, exceeding by far the increase observed for organic C contents and CEC. This suggests that additional mechanisms are in effect. Calcium apparently possesses a high affinity for SOM beyond what would be predicted considering its size and charge.

Calcium is electrostatically bonded to carboxylic groups (- COOH) and is preferred over most other alkaline-earth cations at low concentrations. The mechanism of this preference is not entirely clear. Leenheer et al. found high binding constants for Ca in fulvic acids extracted from the Suwannee River, Georgia. Calcium binding substructures in organic matter that were considered important are the oxy-succinic acid model. Both alkali and alkaline-earth cations are additionally able to form so-called "crown" macrocycles that merely depend on the steric configuration of the cation and the receptor and not on electrostatic bonds. Calcium is selectively complexed in macrocyclic multidentates with a dominant number of N-donors but not in many other macrocycles and ethers. It remains to be shown whether the forms of macrocycles that can act as receptors for Ca are found to a larger extent in ADE. Activated carbons produced from wood, which may have some similarities with incompletely combusted organic C found in the Dark Earths, were shown to have high adsorption capacities for metals, but not specifically for Ca.

The presumably large amount of Ca inputs to these soils can explain the dominance of Ca at the exchange complex, even though this does not explain a particular specificity for Ca. These elaborations are still speculative but may provide suggestions for future research. Phosphorus is mainly present as inorganic phosphate in ADE and may therefore not correlate with SOM contents. These observations may provide an explanation for why, in contrast to P, Ca and C are closely related in ADE. Therefore, the most likely explanation for the observed relationships is based on chemical dynamics in soil and not on different inputs during habitation by Amerindian populations.

Leaching of Nutrients

Many of the Amazonian Dark Earths have a coarse texture, and water percolation is therefore very rapid under the humid tropical climate of Amazônia. Surprisingly, nutrient leaching was very low as determined in a percolation experiment with a Dark Earth from Rio Preto da Eva, north of Manaus, Brazil. This was not an effect of low water percolation, but rather of retention of nutrients in the soil, as water percolation was in fact higher in the ADE than a Ferralsol from the same region.

Cumulative leaching of mineral N, K, Ca, and Mg in the Amazonian Dark Earth was only 24, 45, 79, and 7%, respectively, of that found in a Ferralsol. The large proportion of Ca at the exchange complex explains the relatively high Ca leaching. However, Ca leaching is still very low in comparison to the high availability of Ca in the ADE which led to a 120%

higher Ca uptake compared to an unfertilised control in the same experiment. Nutrient leaching is controlled by the release from adsorption sites and from organic matter by mineralisation, which is apparently sufficiently slow or of low enough intensity to prevent excessive leaching.

When inorganic nutrients were applied to the Amazonian Dark Earths as a fertiliser, however, leaching proved very high and exceeded that found in the Ferralsol. Therefore, the exchange capacity was easily exhausted. This has implications for the management of ADE as well as for the creation of similarly fertile soils in the humid tropics. In unfertilised and unpolluted soils, a large proportion of N, P, and S is usually present in the soil solution in organic forms, as shown for organic N in several watersheds in Chile.

In a Ferralsol under secondary and primary forests near Manaus, Brazil, up to 40-93% of total N and 43-99% of total S in the leachate at 0.1, 0.6 and 2 m depths were contained in dissolved organic matter. These values should be found or even exceeded in ADE, as these possess high SOM contents. The mobility and microbial degradability of dissolved organic nutrients largely depends on whether they are contained in hydrophobic or hydrophilic dissolved organic matter. While S was entirely present as hydrophobic organic matter and therefore less mobile in soil, up to 90% of the total N was found in hydrophilic dissolved organic matter. Studies on leaching in ADE therefore need to include organic nutrient forms to quantify and understand total nutrient losses.

Availability of Soil Moisture

The effect of soil management on soil moisture largely depends on soil texture. Generally, higher SOM contents increase the available water content in soil partly through improved aggregation and partly through the creation of mesopores (0.2-10 μm diameter). Consequently, the field capacity in the topsoil horizons averaged 260 ± 10 mm m^{-1} in ADE and 220 ± 10 mm m-1 in adjacent Ferralsols (B. Glaser, unpubl. data). The effects of higher SOM contents will be more pronounced in soils with a coarse texture than in those with a fine texture, because (i) finely textured soils already have elevated available water contents, and (ii) aggregation of Amazonian Ferralsols is not increased by higher SOM contents. Since many ADE have a coarse soil texture, the organic matter contents play an important role in soil-water relationships. Higher water availability in ADE than in adjacent sites was already mentioned by Hartt reporting greener vegetation during the dry season. On the other hand, aeration is better in ADE than Ferralsols, and water logging and surface runoff can be expected to be lower.

Soil Depth

The depth of the horizons of ADE enriched with organic matter determines the total amount of available nutrients and soil water relationships. Deeper enrichment with SOM increases soil nutrient and water availability, and therefore increases crop production. The total depth of the organically enriched horizons varies substantially between sites. Amazonian Dark Earths can be as shallow as 0.2 m, or can reach to depths of more than 1 m. Total nutrient stocks therefore vary by one order of magnitude between ADE sites. Additionally, depth of enrichment varies within a given site according to the layout of the settlement. In many cases, the greatest enrichment is found not near the topsoil, but at depth.

Production Potential

In an on-farm experiment, maize grain yield and growth were significantly greater on ADE than Ferralsols. However, the variability of grain yield was large and crop yields on ADE with a low production potential were also shown to be lower than those in Ferralsols. The dynamics of maize height revealed more rapid crop growth on ADE than on adjacent soils during initial stages of development, whereas differences diminished towards maturation.

Higher yields on ADE may depend considerably on crop species due to their different nutrient and water requirements. For example, crops with high K requirements, such as bananas, may not grow as well as those crops which have high

P requirements, such as legumes. Consequently bean yields are generally higher on ADE than on unenriched soils. On the other hand, corn production on ADE was not increased compared to that on soils of the periodically inundated Várzea. Manioc was reported to grow more vigorously on ADE of the Tapajos than on adjacent soils with more than 50% increased growth.

Regional and Spatial Variations of Fertility

Costa and Kern studied the geochemical composition of a Dark Earth site in Caxiuanã. The total contents of P, K, Mg, and Ca varied by a factor of 5 within the site. These variations were not associated with random heterogeneity of the soil, but with past human activity, since they show distinct spatial patterns. Already Sombroek mentioned the existence of certain types of ADE (called terra mulata) that did not contain artifacts, but which appeared to be associated with the anthropogenic activity as well as with those Dark Earths that contained artifacts.

McCann et al. found that these soils have distinctly different chemical composition and probably a different genesis, as discussed in Kämpf et al. and Neves et al. They contain similar amounts of organic C, but less available P and Ca). Therefore, different types of ADE at a given location have different fertility and production potentials. Examples of these soils from the Rio Negro and Urubu region that did not contain artifacts but had elevated SOM contents, produced similarly high yields as the ADE with artifacts.

Biological N_2 Fixation

Biological N_2 fixation is the microbial transformation of atmospheric N_2 into organic N. Many microorganisms are able to perform this process, such as Rhizobia in symbiosis with most legumes, or free-living bacteria and blue-green algae. Symbiotic N_2 fixation is by far more effective (3-206 kg N ha-1 yr^{-1} for grain legumes) than the fixation by free-living microorganisms (less than 5 kg N ha^{-1} yr^{-1}). The amount of symbiotically fixed N_2 depends on the soil reaction and the availability of certain nutrients in soil. Low pH, high Al activity and high Mn availability restrict the survival and growth of Rhizobia, as does low P, Ca, Mo and to a lesser extent low Co, B, Fe, Ni, Se, Zn and Cu availability. In addition, high nitrate contents in soil reduce nitrogenase activity and thus N_2 fixation. Amazonian Dark Earths have higher pH and usually no exchangeable Al and lower nitrate or exchangeable Mn contents than adjacent soils, while availability of P, Ca, and Cu is high. These conditions are ideal for maximum biological N_2 fixation. While non-leguminous crops may encounter N deficiency due to low N availablity in Amazonian Dark Earths, grain legumes may not be limited by N availability.

Additionally, legumes may have a competitive advantage in natural regrowth and fallows. Whereas nodules were difficult to find in Ferralsols in a survey comprising multiple locations of the central Amazon Basin, nodulation was abundant in studied ADE of the upper Rio Negro and Novo-Airão. Additionally, 77% of the Amazonian Dark Earths sampled showed positive incidence of Azospirillum spp. compared to only 10% of the Ferralsols. In a secondary forest in Rio Preto da Eva, nodulation of vines was extensive, and the biomass of potentially nodulating woody plants significantly increased in comparison to adjacent non-ADE sites. However, foliar N contents of the legume Vigna unguiculata (L.) Walp. (cowpea), were lower when grown on ADE than on Ferralsol in a greenhouse study, indicating that if biological fixation occurred at all in cowpea, it did not compensate for the lower N availability in the Dark Earth.

References

Castro, C., & Logan, T.J., "Liming effects on the stability and erodability of some Brazilian oxisols," *Soi Science Society of America Journal*, 55, 1407-1413, 1991.

Giller, K.E., *Nitrogen Fixation in Tropical Cropping Systems*, 2nd edition, Wallingford: CAB International, 2001.

LeMare, P.H., Pereira, J., & Goedert, W.J., "Effects of green manure on isotopically exchangeable phosphate in a dark-red latosol in Brazil," *Journal of Soil Science, 38*, 199-209, 1987.

9

Soil Management and Conservation Agriculture

Concerns about soil erosion affecting soil productivity in rainfed areas have resulted in an emphasis on trying to stop negative effects on crop yields attributed to erosion and runoff. This has been attempted by putting cross-slope barriers in fields designed to catch or divert soil and water moving downslope. This approach has not been particularly successful either in halting the problems or in raising yields, resulting in disillusionment among farmers. Money has been spent to little effect and damage to land has not been stopped.

However, if the emphasis is shifted towards the soil as a habitat for roots and if soil loss and runoff are recognised as consequences of prior damage to soil porosity, a different perception emerges. This is based on more positive thinking which considers first the soil conditions that allow plant roots to function optimally, and then the improvements necessary to bring any current inadequate state of the root habitat to that desired condition. Land uses would ideally be cross-matched with variations in land suitability with respect to erosion hazard - i.e. the most protective forms of land use would be allocated to places with the highest hazards of erosion.

However, especially for farmers with few resources and small farms, low yields of subsistence crops may dictate that they be planted on all land units irrespective of erosion hazard. In both situations however, improving soil conditions to meet the needs of plant roots will often greatly reduce problems of soil loss and runoff.

Key goals of improving and maintaining excellent soil conditions for and with roots include:

- — increasing the reliability of plant production in the face of unpredictable variations in the weather and other hazards of the environment;
- — reducing production costs and raising net returns to producers;
- — increasing the quality of the land and its resilience to extreme weather conditions.

Zero Tillage Methods of Crop Production

On increasingly large areas of Latin America there has been a revolution in agricultural practice over the past 30 years. The adoption of zero tillage methods of crop production by large numbers of farmers provides convincing validation of the value of such conservation-effective forms of agriculture, in agronomic, environmental, economic and social terms. This is being achieved on farms whose sizes range from less than twenty hectares to thousands of hectares and in a wide range of ecological zones.

Conservation agriculture (CA), as defined during the First World Congress on Conservation Agriculture (1-5 October 2001) promotes the infiltration of rainwater where it falls and its retention in the soil, as well as a more efficient use of soil water and nutrients leading to higher, more sustainable productivity. It also contributes positively to environmental conservation. In many environments conservation agriculture can be considered the ultimate soil and crop management system. Conservation agriculture has been successfully implemented in both small-scale and large-scale farming, where it has given economic benefits as well as improved water resources.

Zero tillage has been successfully practised in the United States for several decades, with regular annual growth in the total area. In Latin America there has been an impressive rate of adoption and accelerating growth over the past two decades. Brazil and Paraguay suffer erosive rainstorms of very high intensities during the southern summer, which result in severe damage year after year.

On almost all cultivated land, soil tillage for crop production, often with heavy disk ploughs followed by disk harrowing, resulted in many problems. These included:

— loss of the porous organic covering of the forest floor where land had been cleared;
— pulverisation of surface soil together with compaction of the subtillage layer;
— loss of organic matter from the upper soil layers by rapid oxidation from the exposed surface;
— loss of potential soil moisture as runoff;
— reduction in soil depth by erosion of topsoil, resulting in losses of seeds and fertilisers, and causing additional replanting costs;
— declining flow and drying-up of streams and rivers during dry seasons.

Downstream there were problems with eroded sediments clogging urban water purification plants, sedimentation in stream valleys and reservoirs, damage to bridges and roads. A common response was to construct conservation banks on the contour, such as broad- and narrow-based bunds to control runoff and soil erosion. However, they did not stop erosion occurring on uncovered soil. Infiltration of runoff was impeded by the severe compaction along the channel where it collects.

The channel beds are probably the most compact lines in the entire field. As time passed and the runoff and erosion problems continued, larger and larger banks were built, but without conspicuous success in halting the problem. Declining productivity and profitability on family farms resulted in collapsing net farm incomes, falling land prices and families leaving their farms for some other livelihoods.

In 1972 there were 500 hectares under residue-based zero tillage on one farm in southern Brazil. The technique spread slowly at first, because of scepticism and insufficient knowledge. There was a lack of appropriate equipment, suitable cover crops and weed control techniques. As the economic and technical advantages of residue-based zero tillage became apparent the rate of spread accelerated, largely as a result of farmer-to-farmer contacts. By 2001 in Brazil, there were more than 13 million hectares managed in this way. In the State of Santa Catarina, Brazil, residue-based zero tillage has been adopted on 400 000 ha by 1998-1999 within a programme to promote these systems. As a result, some or all of the improved practices were spontaneously adopted on a further 480 000 ha outside the formal remit of the project, from a base of 120 000 ha in 1993-1994.

The State's small farmers have been ingenious in devising their own equipment and methodologies to fit zero tillage to their individual circumstances, together with governmental and non-governmental arrangements for their technical and institutional support. In Paraguay, zero tillage was first used in the late 1970s but was not widely adopted on mechanised medium and large farms until 1990. It had expanded to 20 000 ha by 1993, to 250 000 ha by 1995 - 1996 and to 480 000 ha by 1997, which represents 51 percent of the total cultivated area of Paraguay.

Conservation Agriculture Principles

Planting a crop in the residues of the previous crop, which is the essence of conservation agriculture, is fast becoming a successful and sustainable cropping practice, especially in the subhumid tropics. Implicit in this practice is the absence or limitation of tillage practices that incorporate surface residues or disrupts soil porosity. The quantity of crop residues produced is clearly very important and varies greatly with crop type, variety and yield. Invariably there are residues of weeds associated with crop residues that also contribute to soil cover, especially during the initiation of no-till.

Large quantities of crop residues are usually obtained from sorghum, maize, rice, cotton and sunflower, whereas soybean, wheat and beans generally produce small quantities. Traditional varieties often yield greater quantities of residues than improved varieties, especially those of short stature and high harvest index. Most information on the optimum quantity of crop residues to be left on the soil surface is based on the amounts needed to reduce soil losses to acceptable levels on different slope gradients, rather than the amounts needed to maximise rainwater infiltration.

Data exist showing that cover is less effective in reducing runoff than soil losses, but there is little information on the influence of cover on infiltration and runoff, especially on 20 to 50 percent slopes, which are commonly cultivated by small-scale farmers. Usually, a minimum value of 70 percent surface cover - equivalent to 4-6 t/ha of maise straw for example - should be adopted. The quantity of residues remaining during the cropping season is also influenced by the rate of residue decomposition. Nitrogen-rich legume residues, such as those from beans and soybean, decompose much more rapidly than nitrogen-poor cereal straw and other residues with high C/N ratios.

On the other hand, legumes used as a cover crop can provide a weed-smothering cover, protection from raindrop impact, and important additions to organic matter. Harvesting procedures can drastically affect the quantity of residues remaining in the field. The widely acclaimed success of conservation agriculture is mainly attributed to improved surface porosity that results in increased infiltration and reduced runoff, and a greater water availability to crops.

As additional benefits, conservation agriculture also lessens evaporation losses, reduces erosion, enhances earthworm activity and soil structure, improves soil fertility and lowers labour, machinery and fuel costs. With time, yields increase substantially provided crop rotations are well designed and include leguminous crops or cover crops. When compared with only applying a soil cover (mulches, crop or cover crop residues) in a conventional system, no additional time is required for land preparation in CA (apart from herbicide application in some cases), which allows earlier sowing and all the advantages that this confers. Consequently returns to labour are substantially increased.

There is evidence that the yield of a crop is significantly higher when sown directly into the residues of a previous crop than when it is sown in a previously tilled soil to which the same quantity of crop residues are applied as a mulch. This is attributed to the benefits of little soil disturbance: the soil structure created by the root channels from the previous crops as well as by the biological activity of earthworms and other soil fauna facilitate deeper rooting and enhance the infiltration and percolation of rainwater.

Conservation agriculture principles are implemented optimising the soil as a dynamic habitat for roots as follows:

- Residues of crops and of cover crops are distributed evenly and left on the soil surface.
- Once the soil has been initially brought into good porous condition, no implements are used to turn over the soil, to cultivate it or to incorporate crop residues.
- Weeds and cover crops are controlled by slashing with a knife roller or by preplanting application of a non-polluting desiccant herbicide.
- A specialised planter or drill cuts through the desiccated cover, slotting seed (and fertiliser) into the soil with minimum disturbance.
- Crop rotation is fundamental to zero tillage. It promotes adequate biomass levels for permanent residue cover and assists in control of weeds, pests and diseases. Rotations also ameliorate soil physical conditions, recycle nutrients and can fix atmospheric nitrogen. In semiarid conditions, appropriate crop rotations involving deep-rooting crops can also make still better use of residual soil moisture.
- As a result, soil erosion is reduced by about 90 percent and soil biological diversity maximised

In such systems soil damage is reduced and recuperation of soil architecture is much more quickly achieved than by unimproved fallow systems. Appropriate crop rotations are as important as the soil cover and no-tillage practices. Grasses, in particular, increase the aggregation and stability of soil particles which provide a range of small voids resulting in increased porosity.

Residue-based zero tillage is implemented gradually on structurally damaged soils. At the start, tillage with tined equipment (scarification) can be used to break up the underlying pan and let more rainwater back into the soil, while leaving some of the plant remains on the surface. In this way the soil is opened up and the previous crop's residues are incorporated. It may necessary to start renovating the soil by enabling more rainfall to become soil moisture, but too frequent scarification can also damage soil architecture because of the shattering effect on soil structural units.

Following the break-up of the underlying pan, strip cropping with a legume between rows of the main crop could be carried out. Finally a complete cover of crop residues without further soil disturbance by tillage could be established. The residues change overtime from being a protective cover to becoming an integral component of the soil. In the process, the worms and other soil mesofauna burrow within the soil seeking food and thereby provide channels and biopores through which air and water can move easily.

Effects of Conservation Agriculture

Crop Yields

Farmers' own experiences confirm what was anticipated by results of two six-year experiments with wheat and soybean between 1978 and 1984, comparing effects of conventional tillage, minimum tillage/scarification and zero tillage.

Soil Moisture

It might be expected that zero tillage would be no better than scarification (opening large spaces in the soil and leaving a rough surface) in increasing moisture in the soil, but this is not the case. This shows changes in levels of soil moisture under wheat, at three depths, under conventional soil preparation, scarification (minimum tillage) and zero tillage, during the crop's vegetative stage in the 1981 growing season. Plant-available moisture was greater and water stress, due to drought, shorter under zero tillage than under the other methods.

The fact that differences in the three-dimensional arrangement of the root habitat contribute to differences in root growth and function, even though soil moisture conditions may be almost the same, has profound implications. The best conditions for root growth and function appear to be where there has been no disturbance by tillage implements and where soil organisms are doing the work of burrowing, transforming and aggregating soil constituents.

It may also be that differences in soil moisture inferred from runoff measurements under different tillage treatments may be insufficient to explain differences in root measurements and in final yield. A pioneer farmer in Paraná, Brazil, whose soil conditions have been monitored from 1978 to the present, has kept detailed yield records. These show that under residue-based zero tillage, yields of both maize and soybean have been rising and have become less variable from year to year.

Effects on Other Soil Health Indicators

The impacts of zero tillage (ZT) and conventional tillage (CT) on soil health are shown by comparing some soil indicators for both systems:

- — diameter and stability of soil aggregates
- — soil organic matter content at 20 cm depth
- — number of earthworms

Saturnino and Landers measured the number of maize roots in each 10 cm layer of soil to 1 m depth after 15 years of constant treatment (zero tillage and conventional tillage). The results in Table 1 show marked differences.

Table 1. Buildup of soil organic matter under ZT compared with conventional cultivation

System and duration	Mean organic matter* 0-20 cm depth (%)
CT	2.5
Zero tillage - 4 years	2.7
Zero tillage - 7 years	2.9
Zero tillage - 10 years	3.1

* Discounting the crop residue mulch layer above

Zero tillage and crop rotation favour recycling of nutrients and better soil structure, resulting in better root development and higher production. A research report from 1983 showed similar differences in root distribution of soybeans. While the total number of roots was the same to 1 m depth, they were more evenly distributed down the profile with zero tillage than with conventional tillage.

Erosion and Runoff

Conservation agriculture compared with conventional tillage results in markedly reduced soil erosion and runoff, as shown in results from Brazil and Paraguay. This effect is attributed to the increased soil porosity beneath residues due to biological activity. Note the saving of 441 mm water by reduction of runoff in southern Brazil and 186 mm in central Brazil. In the Municipio of Tupanssi in Paraná it was reported that, following adoption of residue-based zero tillage, the turbidity of the river water has fallen from an index of 8 000 to 80 (author's field notes).

A group of farm families, whose houses were on the slopes of cultivated fields recently transformed by zero tillage, said they were pleased that the runoff water and sediment no longer rushed down the hillsides into their houses, damaging the rugs and carpets on the floors (author's field notes). If runoff and erosion are symptoms of soil misuse, the major reduction in both occurrences signifies that their causes must have been significantly reduced.

Catchment Hydrology

An example of positive changes in catchment hydrology is provided by a representative catchment near Toledo in Paraná, Brazil. Soon after the adoption of zero tillage on rolling wheat lands, farm families observed that a pond which formerly had been dry for much of the year filled with water and hydrophytic vegetation took hold again.

Further down the catchment, the river, which had ceased to flow in the dry season began to flow again throughout the year, so that a small farmer on its banks was able to improve his livelihood by investing in irrigation equipment and in excavating fishponds. This farmer now keeps fishponds full of water through the year and charges people to come fishing for fun at the weekends.

Zero Tillage Systems on Farm Economics

Farmers have responded to the economic benefits of zero tillage. Yield increases of 20 percent or more, coupled with reduction of production costs by a similar percentage, have had positive effects on farm income. Savings of time and labour have contributed to improvements in farm families' livelihoods. For instance in Paraguay, on farms using conventional tillage systems, severe losses of soil, nutrients and organic matter were seen as a root cause of declining yields of a range of crops. Some farms had adopted zero tillage, others not.

Farm records over 10 years were used to construct economic models and indicators of differences. On representative mechanised 135 ha farms growing rotations including oats, soybean, sunflower, maize, wheat, crotalaria, vetch with zero tillage (ZT) farm incomes rose while those using conventional tillage (CT) for rotations with soybean, oats, wheat, maize fell. The returns on capital increased on farms using zero tillage, but declined on those using conventional tillage.

Reduction of tractor-hours, reduced use of fuel and lower costs of repairs, etc. contributed to the economic benefits of zero tillage on these farms. In another study in Paraguay, the economics of zero tillage on seven smaller farms (20 ha or less) without tractors were studied. Five out of the seven farmers had both conventional and zero tillage areas on their properties.

The small-farm study illustrates that zero tillage is not only financially attractive to small farmers but also has high economic pay-off for the nation. In Paraguay it has been estimated that for 1997 the national economic benefit due to the adoption of zero-tillage systems reached US$941 million. These included the saving in nutrients lost from soil from erosion, plus the costs saved in reduced tractor hours, less fuel and fertiliser. The 1980 agricultural census of Paraná State, Brazil showed that there were over 6 million ha of annual crops. A 1989 report indicated the annual benefits if residue-based zero tillage systems were to be applied to the full 6 million ha.

Residue-based Zero Tillage Systems in Latin America

From the full application of both the concepts and integrated techniques of residue-based zero tillage (also called Conservation Agriculture), farmers have achieved many direct and indirect benefits, often recorded together on individual farms.

On-farm benefits included:

- marked and rapid increase of organic matter content in upper layers of soil and increased biodiversity, number and activity (of earthworms, fungi, bacteria, etc.) in the soil;
- better soil structure and stability of soil aggregates; significantly higher infiltration rates; soil loss reduced by over 80 percent, runoff by 50 percent or more; more intensive but safe use of sloping areas made possible;

- increase in nutrients stored, greater availability of P, K, Ca, Mg in the root zone; less fertiliser needed for same result;
- better germination and development of plants, better root development and to much greater depth; better resilience of crops in rainless periods due to increased water holding capacity;
- yields often higher, typically + 20 percent for maize, + 37 percent for beans, + 27 percent for soybean, + 26 percent for onions; with less year-to-year yield variation;
- reduced variations of soil temperature during the day, with positive effects on plants' absorption of water and nutrients;
- less investment and reduced use of machinery and animals in crop production; reduced costs for labour, fuel and machinery-hours perceptible within 2 years. Operational net margins per ha rose by between + 58 percent and + 164 percent, because of combination of lower cost of production and increase in yields, which provides greater resilience against falling market prices and bad weather;
- greater flexibility in farm operations especially over optimum dates for planting; increasing possibilities for diversification into livestock, high-value and different crops, vertical integration into product processing and other activities; improved quality of life.

Off-farm benefits widely noted by rural agency staff and others, included:

- flooding risks reduced by 30-60 percent due to greater rainfall infiltration and delays to overland flows. Extending the time of concentration; better recharge of underground aquifers, improving groundwater reserves and dry season flow in springs and streams;
- less herbicide use after first years; less pesticide use, more recycling of animal wastes; reduction of pollution and eutrophication of surface waters by agricultural

chemicals carried in surface runoff and eroded soil; less sedimentation and infrastructure damage, e.g. silting of waterways, large dams. A conservative estimate for the Cerrado region was given as US$33 million per year;

— reduced water treatment costs (ca. 50 percent) due to less sediment, less bacterial and chemical contamination;

— savings of up to 50 percent in costs of maintenance and erosion avoidance on rural roads;

— reductions in fuel consumption of 50-70 percent or more and proportional reduction in greenhouse gas emissions;

— reduced pressure on the agricultural frontier and reduced deforestation by high-yielding, sustainable conservation agriculture and increased pasture carrying capacity through rotation with annual crops;

— enhanced diversity and activity of soil biota;

— reduced carbon emissions through less fuel use and enhanced carbon sequestration by not destroying crop residues and increasing, rather than losing, soil organic matter.

The zero tillage systems of Latin America thus are not only a great improvement on former tillage-based systems, but also have major off-site and national benefits, to which improvements in soil moisture management make a large contribution. The effects are illustrated by the colour of the water going over the Iguassu Falls in southern Brazil. By chance these two Plates were taken from the same viewpoint 7 years apart, one in the wet season when high runoff also transported much eroded soil, the other in the dry season when water that had seeped down through the soil to the groundwater provided the dry-season flow.

Constraints of CA

Conservation agriculture has been successfully employed in subhumid as well as humid climates, but there are still some constraints in semiarid environments that may hinder its immediate application. Typical of these constraints are:

— shortage of water limiting crop and residue production;
— insufficient residues produced by the economically or socially important crops and lack of knowledge of suitable cover crops;
— sale or preferential use of crop residues for fodder, fuel and building materials;
— inability to control livestock grazing, especially in areas where communal grazing is traditional (tenant farmers are often obliged to allow the landowner's cattle to graze the residues after harvest);
— inability to control residue consumption by termites;
— insufficient money or credit to purchase appropriate equipment and supplies;
— lack of knowledge of conservation agriculture by extension and research staff.

A number of approaches have been explored and are being tested to overcome these constraints. In situations where crop residues are preferentially used as fodder, additional new sources of fodder may be produced, provided they can be protected from grazing by, for example, live fences. Hay or silage may be produced as additional dry-season fodder from improved pasture species, or from forage trees or crops of high biomass grown specifically for this purpose.

Forage trees can be established as live fences along farm and field boundaries, and forage grasses may be produced as live barriers, on bunds, and along field boundaries and roadways. In Bahir Dar, Ethiopia, farmers are increasing fodder production by undersowing forage legumes in other crops, establishing forage strips between arable crops, and by oversowing mixtures of legume seeds on grazing areas. Certain crop sequences are less suited to direct sowing into crop residues because of the likelihood that weed, pest or disease problems will become intensified by being transmitted from one crop to the next. Examples of less suitable crop sequences and their specific problems encountered in eastern Bolivia are:

- — wheat every year - disease problems;
- — soybean every year - pest and disease problems;
- — soybean-sunflower sequences - disease problems;
- — maize-sorghum or sorghum-black oats - weed and pest problems;
- — sunflower-cotton - the problem of volunteer sunflower weeds;
- — bean-soybean sequences - pest and disease problems.

Weed problems may also be caused by volunteer germination of the previous crop; for example, sunflower volunteers can be particularly difficult to eradicate. To avoid such problems, appropriate crop rotations, acceptable to the farmers, must be selected. In environments where there are many constraints to the introduction of conservation agriculture, a pragmatic, phased approach may be the most feasible, in which individual constraints are progressively overcome until an appropriate system of conservation agriculture can be fully implemented.

This may require the planned introduction of measures such as improved grass species and fodder trees, hay and silage production, live fences, stall-fed livestock, improved crop rotations with cover crops, formation of farmers' associations, credit supply and local or international training visits for farmers, extension and research staff.

The introduction of conservation agriculture is unlikely to be immediately successful on seriously degraded soils with surface crusts, compacted layers, low fertility or severe weed infestations unless these problems are first overcome by appropriate remedial actions. Hardsetting soils may not be immediately suitable for conservation agriculture because of the difficulties of overcoming soil compaction problems and maintaining good soil porosity within the topsoil and subsoil.

Consequently crop rooting is frequently restricted to shallow depths. In this case, deep tillage followed by the establishment of cover crops prior to introducing conservation agriculture, and then the adoption of crop rotations that produce large quantities of residues, will progressively improve the physical condition of these soils and make conservation agriculture possible. Conservation agriculture is less likely to be successful in poorly drained soils because the added residues will intensify anaerobic conditions, in which toxic substances harmful to crop growth may be produced.

The cost of no-till planters and seed drills needed for direct sowing may be a major constraint for mechanised farmers, unless it is possible to modify their existing seed drills and planters. For small farmers, hand tools and animal-drawn equipment exist and local blacksmiths can often adapt them, provided they have access to information and samples.

References

Haas, H.J., Evans, C.E., and Miles, E.F., *Nitrogen and Carbon Changes in Great Plains Soils as Influenced by Cropping and Soil Treatments*, USDA, Washington, DC, Technical Bulletin 1164, 1957.

Paul, E.A. and Clark, F.E., *Soil Microbiology and Biochemistry*, 2nd ed., Academic Press, San Diego, CA, 1996.

Rasmussen, P.E. *et al.*, "Crop residue influences on soil carbon and nitrogen in a wheat-fallow system," *Soil Sci. Soc. Am. J.*, 44, 596–600, 1980.

Waldrop, M.P., Zak, D.R., and Sinsabaugh, R.L., "Microbial community response to nitrogen deposition in northern forest ecosystems," *Soil Biol. Biochem.*, 36, 1443–1451, 2004.

10

Soil Remediation Technologies

There are two factors that impact on the cost-effectiveness of remediation technologies. The first is the impact of waste legislation and regulation that, in certain nations, determines the fate of contaminated soil, and the potential for its treatment, disposal, recovery, recycling and reuse. The second is the designated land-use of a remediated site; this has a profound effect on site values and hence the options available for remediation.

In general remedial options fall into one or more of the following broad categories:

— Excavation and containment (Removal to landfill: the disposal of material to an engineered commercial void space; Deposition within an on-site engineered cell, generally with a view to combining the disposal of waste with the reclamation of land area from the void space; Engineered land-raising and land forming, where materials are deposited on the land surface to make a hill or mound above the natural surface level suitably contained.)

— Engineered systems (In situ Physical Containment: designed to prevent or limit the migration of contaminants left in place or confined to a specific storage area, into the wider environment. Approaches include in-ground barriers, capping and cover systems; Hydraulic containment and pump-to-contain approaches.)

— Site rehabilitation measures are those used to bring back some measure of utility to a site whose contamination cannot be treated or contained for technical or economic reasons. Examples include growth of grass cover tolerant of contaminants, covering with soil or soil substitute, liming and other cultivation measures.

— Treatment based approaches destroy, remove or detoxify the contaminants contained in the polluted material. Using treatment technologies in contaminated land remediation is encouraged by agencies in many countries, because they are perceived as having added environmental value compared with other approaches to remediation such as excavation and removal, containment or covering / revegetation. The "added" environmental value is associated with the destruction, removal or transformation of contaminants into less toxic forms.

Treatment based approaches can be described as:

— *Biological Processes (Bio)*: contingent on the use of living organisms

— *Chemical processes (Chem)*: destroy, fix or concentrate toxic compounds by using one or more types of chemical reaction

— *Physical processes (Phys)*: separate contaminants from the soil matrix by exploiting physical differences between the soil and contaminant (e.g. volatility) or between contaminated and uncontaminated soil particles (e.g. density).

— *Solidification and stabilisation (S/S)*: processes immobilise contaminants through physical and chemical processes (Solidification processes are those which convert materials into a consolidated mass. Stabilisation processes are those in which the chemical form of substances of interest is converted to a form which is less available).

— *Thermal processes*: exploit physical and chemical processes occurring at elevated temperatures.

Ex situ approaches are applied to excavated soil and/or extracted groundwater. In situ approaches use processes occurring in unexcavated soil, which remains relatively undisturbed. On site techniques are those that take place on the contaminated site. They may be ex situ or in situ. Off site processes treat materials that have been removed from the excavated site (ex situ).

Selecting Effective Soil Remedial Actions

There is a number of factors that need to be considered in selecting an effective remediation solution. These include considerations of core objectives such as risk management, technical practicability, feasibility, cost/benefit ratio and wider environmental, social and economic impacts. In addition, it is also important to consider the manner in which a decision is reached. This should be a balanced and systematic process founded on the principles of transparency and inclusive decision-making. Decisions about which risk management option(s) are most appropriate for a particular site need to be considered in a holistic manner. Key factors in decision making include:

- — Driving forces to remediate and goals for the remediation objectives;
- — Risk management;
- — Sustainable development;
- — Stakeholders' views;
- — Cost effectiveness
- — Technical feasibility

Driving Forces and Goals for the Remediation Objectives

Most remediation work has been initiated for one or more of the following reasons:

- — *To protect human health and the environment*: In most countries, legislation requires the remediation of land, which poses significant risks to human health or other receptors in the environment such as groundwater or surface water. The contamination could either be from "historic" contamination or recent spillage of substances from a process or during transport. Groundwater protection has in many countries become an important driver for remediation projects.

— *To enable redevelopment*: Remediation of formerly used land may take place for strictly commercial reasons, or because economic instruments have been put in place to support the regeneration of a particular area or region; and/or (Brownfields redevelopment concept).
— *To repair problems*: In some cases remediation work must be retrofitted to a newly developed site.
— *To limit potential liabilities*: Remediation can take place as an investment to increase the potential value of land. Owners may perceive that a particular site could potentially have an environmental impact, which might leave them liable to third party actions.

Risk -based Approach

A risk-based approach has been adopted for the management of contaminated land in many countries. The assessment and management of land contamination risks involves three main components:

— The source of contamination;
— The receptor; and
— The pathway (the route by which a receptor could come into contact with the contaminating substances).

A pollutant linkage exists only when all three elements are in place. The probability that a pollutant linkage exists needs to be assessed. Risk assessment involves the characterisation of such a relationship, which typically includes: delineation of the source, measurement and modelling of fate and transport processes along the pathway, and assessment of the potential effect on and behaviour of the receptor.

A consideration of risk must also take into account of not only the existing situation but also the likelihood of any changes in the relationship into the future. From a risk management standpoint, remediation technologies are applied to the control of the source term and/or the management of contaminants along the pathway.

Concept of Sustainable Development

The concept of sustainable development gained international governmental recognition at the United Nation's Earth Summit conference in Rio de Janeiro in 1992. A number of definitions for sustainable development have been proposed in different countries, based on the original consept of: "Development that meets the needs of the present without compromising the ability of future generations to meet their own needs". Underpinning all of these approaches are three basic elements to sustainable development: economic growth, environmental protection and social progress.

At a strategic level, the remediation of contaminated sites supports the goal of sustainable development by helping to conserve land as a resource, preventing the spread of pollution to air, soil and water, and reducing the pressure for development on Greenfield sites. However, remediation activities themselves have their own environmental, social and economic impacts. On a project-by-project basis, the negative impacts of remediation should not exceed the benefits of the project.

At present there are no generally agreed means of carrying out sustainability appraisal for remediation projects. Although approaches to assessing the wider impacts of individual elements of sustainability are under development in several countries, a truly integrated approach has yet to be found. There is some way to go before an international consensus can be reached in the way that agreement has emerged about the principles of risk assessment and risk management. This is hardly surprising given the complex interplay of economic, environmental and social factors that affect and are affected by a remediation project.

Remediation objectives typically relate to environmental and health risks and perhaps performance of geotechnical / construction measures. These may form part of a larger regeneration project with social and economic aims, such as attracting inward investment. What is realisable, and the approaches that can be taken, will be subject to certain site/ project specific boundaries, for example the time and money available for the remediation works, the nature of the contamination and ground conditions, the site location and many more.

Hence the objectives that can be realised by remediation works represent a compromise between desired environmental quality objectives and these site-specific boundaries. This compromise is reached by a decision making process involving several stakeholders. This decision making process is often protracted and costly. Its conclusions can be said to represent the core of the remediation project. While achieving environmental quality objectives will normally underpin any project dealing with contaminated land, desired quality objectives may be driven by a combination of technical criteria and third party nontechnical perception of risk.

From a broader perspective remediation processes will achieve these core objectives, by:

— Helping to conserve land as a resource

— Preventing the spread of pollution to air and water

— Reducing the pressure for development on green field sites

In a broader perspective, a number of questions must also be addressed, such as:

— How justified is it to excavate ground materials and put them somewhere else?

— How justified is it to burn huge amounts of fossil fuel to treat ground containing grams of hydrocarbon per kilogram?

— How justified is it to strip VOC to atmosphere?

The grand question is therefore, should the remedial regime also be sustainable? If the undesirable impacts of these remediation processes exceed the desired benefits of the core objectives, the core objectives may need to be re-evaluated. If proper risk management procedures have been followed, along with a thorough cost benefit analysis and stakeholder consultation, the risks of such a situation arising should be minimised, depending on the remediation approach selected.

Different remediation approaches will vary in their wider environmental impacts, and perhaps also their wider social and economic effects. For example, the acceptability to local residents of different processes can differ. It is therefore useful to consider the route taken to affect the remediation, as well as the core objectives of the remediation project. Assuming an overall "sustainability value" of the core objectives these "non-core" considerations help determine the remediation approach, which detracts least from this overall value.

The wider consequences of a particular remedial project are site-specific in their nature. Some may be temporary, other permanent. The significance of these consequences also depends on the location of the site being remedied. The importance of "nuisance" issues associated with remedial options, may, for example be less for a remote site than for a site in a city neighbourhood. The relative significance that attaches to any particular wider effect of remediation will itself vary at a local, regional and / or national level, for example as a result of cultural differences, differences in population density, use of resources etc.

Stakeholders View in Remediation

The principal stakeholders in remediation are generally considered to be the "problem owners" (usually the polluter or site owner), and also all those with an interest in the land, its redevelopment, and the environmental, social and financial impacts of any necessary risk management works. Depending on the size and prominence of the site these stakeholders will include several of the following:

— Land owners
— Problem holders
— Regulatory authorities
— Planning authorities
— Site users, workers, visitors
— Financial community (banks, founders, lenders, insurers)
— Site neighbours (tenants, dwellers, visitors)
— Campaigning organisations and local pressure groups

Consultants, contractors, and possibly researchers Stakeholders will have their own perspective, priorities, concerns and ambitions regarding a site. The most appropriate remedial actions will offer a balance between meeting as many of their needs as possible, in particular risk management and achieving sustainable development, without unfairly disadvantaging any individual stakeholder. It is worth noting at this point that for some stakeholders, the end conditions of the site are likely to be significantly more important than the actual process used to arrive at that condition. Such actions are more likely to be selected where the decision-making process is open, balanced, and systematic.

Given the range of stakeholder interests, agreement of project objectives and project constraints such as use of time, money and space, can be a time consuming and expensive process. Seeking consensus between the different stakeholders of a decision is important in helping to achieve sustainable development. Risk communication and risk perception issues need special considerations. Arguments and decisions need to be communicated in a balanced form to all stakeholders.

A diverse range of stakeholders may need to reach agreement before specific remedial objectives can be set, for example, site owner, regulators, planners, consultants, contractors, site neighbours and perhaps others. Unsurprisingly once these remedial objectives are set it may be hard to renegotiate them. In most practical situations, the members of the decision making team who are finally responsible for the choices of technology are the landowner, the funder, the regulator and the service provider. All other stakeholders are in a position of influence but in most cases their input does not control the decision. Landowners provide finance to execute projects. Regulators ensure compliance with acceptable environmental quality standards. Service providers apply their expertise to deliver results for both parties.

Costs of Remediation

Costs of remediation depend on many factors and may be broken down into mobilisation, operation (per unit volume or area treated), demobilisation, monitoring and verification of performance. Although data can only be tentative, comparisons of indicative equivalent costs may be a useful exercise in the early stages of consideration of different remediation options.

A significant amount of remediation takes place as a result of the redevelopment of brownfield sites, either as a private commercial venture or as part of a wider regeneration initiative often supported by public funds. Brownfields remediation is discussed as a specific topic in the Final Report of CLARINET. Typically brownfields regeneration aims at stimulating wider economic regeneration by the attraction of new industries or other commercial activities. However, regeneration projects can also be suitable for "softer" end uses, for example "country parks" in areas where the commercial drivers for redevelopment are less.

The future use of land, and the money available for developing this use, is powerful controlling influences on the nature of remediation technologies that can be used. There is a constant pressure for lower remediation costs, both to improve the economics of brownfield reuse for "hard applications" such as housing or commerce; and for "softer" uses such as non-food (energy crops) agriculture. There is growing pressure to develop more cost-effective remediation technologies.

Cost effectiveness is not just a product of reducing remediation costs, but also of finding remediation approaches that provide an additional enhancement to the value of the land. There are two further factors that impact on the cost-effectiveness of remediation technologies. The first is the impact of waste legislation and regulation that, in certain nations, determines the fate of contaminated soil, and the potential for its treatment, disposal, recovery, recycling and reuse. The second is the designated land-use of a remediated site; this has a profound effect on site values and hence the options available for remediation.

Technical and Environmental Criteria

Remedial approaches can be categorised in a way that makes it easier to compare their suitability in general for particular problems, and their feasibility for more specific site circumstances. A suitable technology is one that meets the technical and environmental criteria for dealing with a particular remediation problem. However, it is also possible that a proposed solution may appear suitable, but is still not considered feasible, because of concerns about:

- Previous performance of the technology in dealing with a particular risk management problem (in the countries);
- Ability to offer validated performance information from previous projects;
- Expertise of the purveyor;
- Ability to verify the effectiveness of the solution when it is applied;
- Confidence of stakeholders in the solution;
- Cost; and
- Acceptability of the solution to stakeholders who may have expressed preferences for a favoured solution or have different perceptions and expertise.

In general, concerns over feasibility tend to be greater for innovative remedial approaches, even if these have long standing track records in other countries. However, it is often these innovative solutions that are seen to offer more in terms of reducing wider environmental impacts and furthering the cause of sustainable development.

Types of Soil Remediation Technologies

Soil remediation technologies can be categorised in three broad groups according to the fundamental principles involved in the contaminant removal or treatment process.

Mass Transfer Technologies

This group includes technologies that remove contaminant mass from the soil matrix by physical or chemical means, and subsequently treat or destroy them in other process step. Such removal may or may not involve a change in the phase of the contaminant. Strictly speaking, because these technologies always invlove transfer mass from soil solids, a change in phase will always occur. This group includes technologies that can be applied either in situ or ex situ, such as soil vapour extraction, low temperature thermal desorption and solvent extraction.

Transformation Technologies

This group includes technologies that transform the contaminant mass into products of different chemical compositions by various chemical or biochemical means. The purpose of these technologies is to transform the contaminant into harmless by-products or into a new form that is easier to treat or dispose. Such technologies include bioremediation and thermal destruction.

Stabilisation Technologies

This group includes technologies that incorporate the contaminants into a solid matrix so that leaching into the environment is reduced to levels below those required by regulatory agencies. The incorporation of the contaminant into a monolithic structure can be accomplishes by physical or chemical means or by a combination of both. Such technologies include cement or lime stabilisation, vitrification and other macro or microencapsulation techniques.

Treatability Test and Remediation Technology

From a technical standpoint, a remedial plan generally consists of four distinct phases:

Technology Screening and Selection

The objective of this phase of the study is to identify, in a relative short time, potentially effective technologies as candidates for bench scale testing. Target clean up values are generally the most important criteria for choosing potential treatment technologies. Cleanup levels must be clearly defined before beginning the selection and evaluation process.

The fist step in this phase I requires site and soil characterisation. The site hydrogeologic characteristic and soil properties must be determined and the extent and level of contamination defined. The second step involves information studies and data collection related to the contaminant type, site characteristics, and various treatment technologies. Information can be obtained from many remedial technology investigations already implemented at industrial scale.

In step 3, lab scale (test tube/shaker flask) testing may be conducted as part of the information gathering effort. These test are relatively fast and inexpensive and can provide valuable guidance for technology selection. Carefully planned and executed lab experiments can be part of the initial technology screening an evaluation. The may also be performed to determine key parameters such as biodegradability of contaminants and sorption characteristics of the soil.

Step 4 assesses the need for pre-treatment and or post treatment of soil and other pollutant streams. For instance soil pre-treatment may be required to remove debris and other large particles and to minimise the effects of soil clodding, which may have a profound effect on the removal efficiencies of mass transfer processes Once all pertinent information is gathered and evaluated a decision is reached concerning which technologies will be tested at bench scale level. This is accomplished in steps 5 and 6. If more than one process is selected, as may be necessary in many cases, a testing protocol must be developed for each selected technology.

Bench-scale Testing

The primary objective of bench-scale testing is to assess the effectiveness of a given remedial technology and generate data that will allow estimation of pertinent design parameters. A bench scale-scale study should produce reliable information at a low cost and in relatively short time.. The average duration for bench scale testing of physicochemical technology is approximately four months.

The bench scale protocol should be flexible and allow exploration of new directions, provided that experimental findings obtained during its execution support such a deviations. The bench scale protocol should provide details and step by step instructions of the following:

- Materials and supplies
- Construction and equipment needed
- Sample handling and preparation
- Preparation of required solutions
- Experimental QA/QC plan (duplicates, control..)
- Scheduling and execution of the experimental procedures
- Sampling programme
- Development of analytical methods and analytical QA/QC
- Evaluation of technology effectiveness
- Estimation of scale-up parameters.

Physico-Chemical Remediation Technologies

Soil Washing Systems

Contaminants sorbed onto fine soil particles are separated

from bulk soil in an aqueous-based system on the basis of particle size. The wash water may be augmented with a basic leaching agent, surfactant, pH adjustment, or chelating agent to help remove organics and heavy metals.

Ex situ soil separation processes (often referred to as "soil washing"), mostly based on mineral processing techniques, are widely used in Northern Europe and America for the treatment of contaminated soil. Soil washing is a water-based process for scrubbing soils ex situ to remove contaminants. The process removes contaminants from soils in one of the following two ways:

- — By dissolving or suspending them in the wash solution (which can be sustained by chemical manipulation of pH for a period of time); or
- — By concentrating them into a smaller volume of soil through particle size separation, gravity separation, and attrition scrubbing (similar to those techniques used in sand and gravel operations).

Soil washing systems incorporating most of the removal techniques offer the greatest promise for application to soils contaminated with a wide variety of heavy metal, radionuclides, and organic contaminants. Commercialisation of the process, however, is not yet extensive. The concept of reducing soil contamination through the use of particle size separation is based on the finding that most organic and inorganic contaminants tend to bind, either chemically or physically, to clay, silt, and organic soil particles.

The silt and clay, in turn, are attached to sand and gravel particles by physical processes, primarily compaction and adhesion. Washing processes that separate the fine (small) clay and silt particles from the coarser sand and gravel soil particles effectively separate and concentrate the contaminants into a smaller volume of soil that can be further treated or disposed of. Gravity separation is effective for removing high or low specific gravity particles such as heavy metalcontaining compounds.

Attrition scrubbing removes adherent contaminant films from coarser particles. However, attrition washing can increase the fines in soils processed. The clean, larger fraction can be returned to the site for continued use. Complex mixture of contaminants in the soil (such as a mixture of metals, nonvolatile organics, and SVOCs) and heterogeneous contaminant compositions throughout the soil mixture make it difficult to formulate a single suitable washing solution that will consistently and reliably remove all of the different types of contaminants. For these cases, sequential washing, using different wash formulations and/or different soil to wash fluid ratios, may be required.

Soil washing is generally considered a media transfer technology. The contaminated water generated from soil washing are treated with the technology(s) suitable for the contaminants. The duration of soil washing is typically short- to medium-term. Factors that may limit the applicability and effectiveness of the process include:

— Complex waste mixtures make formulating washing fluid difficult.

— High humic content in soil may require pretreatment.

— The aqueous stream will require treatment at demobilisation.

— Additional treatment steps may be required to address hazardous levels of washing solvent remaining in the treated residuals.

— It may be difficult to remove organics adsorbed onto clay-size particles.

Chemical Extraction

Chemical extraction does not destroy wastes but is a means of separating hazardous contaminants from soils, sludges, and sediments, thereby reducing the volume of the hazardous waste that must be treated. The technology uses an Extracting chemical and differs from soil washing, which generally uses water or water with wash-improving additives. Commercial-scale units are in operation. They vary in regard to the

Chemical employed, type of equipment used, and mode of operation. Physical separation steps are often used before chemical extraction to grade the soil into coarse and fine fractions, with the assumption that the fines contain most of the contamination. Physical separation can also enhance the kinetics of extraction by separating out particulate heavy metals, if these are present in the soil.

Acid Extraction

Acid can also be used as the extractant. Acid extraction uses hydrochloric acid to extract heavy metal contaminants from soils. In this process, soils are first screened to remove coarse solids. Hydrochloric acid is then introduced into the soil in the extraction unit. The residence time in the unit varies depending on the soil type, contaminants, and contaminant concentrations, but generally ranges between 10 and 40 minutes.

The soil-extractant mixture is continuously pumped out of the mixing tank, and the soil and extractant are separated using hydrocyclones. When extraction is complete, the solids are transferred to the rinse system. The soils are rinsed with water to remove entrained acid and metals.

The extraction solution and rinse waters are regenerated using comercially available precipitants, such as sodium hydroxide, lime, or other proprietary formulations, along with a flocculent that removes the metals and reforms the acid. The heavy metals are concentrated in a form potentially suitable for recovery. During the final step, the soils are dewatered and mixed with lime and fertiliser to neutralise any residual acid.

Solvent Extraction

Solvent extraction is a common form of chemical extraction using organic solvent as the extractant. It is commonly used in combination with other technologies, such as solidification/ stabilisation, incineration, or soil washing, depending upon site-specific conditions. Solvent extraction also can be used as a stand alone technology in some instances. Organically bound metals can be extracted along with the target organic contaminants, thereby creating residuals with special handling requirements.

Traces of solvent may remain within the treated soil matrix, so the toxicity of the solvent is an important consideration. The treated media are usually returned to the site after having met Best Demonstrated Available Technology (BDAT) and other standards. The duration of operations and maintenance for chemical extraction is medium-term

Stabilisation and Solidification Technologies

Contaminants are physically bound or enclosed within a stabilised mass (solidification), or chemical reactions are induced between the stabilising agent and contaminants to reduce their mobility (stabilisation). S/S contaminants are physically bound or enclosed within a stabilised mass (solidification), or chemical reactions are induced between the stabilising agent and contaminants to reduce their mobility (stabilisation). Ex situ S/S, however, typically requires disposal of the resultant materials. Sometimes, material can be replaced on site. There are many innovations in the stabilisation and solidification technology.

Most of the innovations are modifications of proven processes and are directed to encapsulation or immobilising the harmful constituents and involve processing of the waste or contaminated soil. Nine distinct innovative processes or groups of processes include:

- bituminisation,
- emulsified asphalt,
- modified sulfur cement,
- polyethylene extrusion,
- pozzolan/Portland cement,
- radioactive waste solidification,
- sludge stabilisation,
- soluble phosphates, and

— vitrification/molten glass.

Typical ex situ S/S is a short- to medium-term technology.

Modified Sulfur Cement

Modified sulfur cement is a commercially-available thermoplastic material. It is easily melted (127° to 149° C (260° to 300° F)) and then mixed with the waste to form a homogenous molten slurry which is discharged to suitable containers for cooling, storage, and disposal. A variety of common mixing devices, such as, paddle mixers and pug mills, can be used. The relatively low temperatures used limit emissions of sulfur dioxide and hydrogen sulfide to allowable threshold values.

Pozzolan/Portland Cement Process

Pozzolan/Portland cement process consists primarily of silicates from pozzolanic-based materials like fly ash, kiln dust, pumice, or blast furnace slag and cement-based materials like Portland cement. These materials chemically react with water to form a solid cementious matrix which improves the handling and physical characteristics of the waste. They also raise the pH of the water which may help precipatate and immobilise some heavy metal contaminants.

Pozzolanic and cementbased binding agents are typically appropriate for inorganic contaminants. The effectiveness of this binding agent with organic contaminants varies. The target contaminant group for ex situ S/S is inorganics, including radionuclides. Most S/S technologies have limited effectiveness against organics and pesticides, except vitrification which destroys most organic contaminants Factors that may limit the applicability and effectiveness of the process include:

- Environmental conditions may affect the long-term immobilisation of contaminants.
- Some processes result in a significant increase in volume (up double the original volume).

- — Certain wastes are incompatible with different processes.
- — Treatability studies are generally required.
- — Organics are generally not immobilised.
- — Long-term effectiveness has not been demonstrated for many contaminant/process combinations.

Incineration

High temperatures, 870-1,200 °C (1,600- 2,200 °F), are used to combust (in the presence of oxygen) organic constituents in hazardous wastes. High temperatures, 870 to 1,200 °C (1,400 to 2,200 °F), are used to volatilise and combust (in the presence of oxygen) halogenated and other refractory organics in hazardous wastes.

Often auxiliary fuels are employed to initiate and sustain combustion. The destruction and removal efficiency (DRE) for properly operated incinerators exceeds the 99.99% requirement for hazardous waste and can be operated to meet the 99.9999% requirement for PCBs and dioxins. Off gases and combustion residuals generally require treatment.

Circulating Bed Combustor (CBC)

Circulating bed combustor (CBC) uses high velocity air to entrain circulating solids and create a highly turbulent combustion zone that destroys toxic hydrocarbons. The CBC operates at lower temperatures than conventional incinerators (1,450 to 1,600 °F). The CBC's high turbulence produces a uniform temperature around the combustion chamber and hot cyclone. The CBC also completely mixes the waste material during combustion.

Infrared Combustion

The infrared combustion technology is a mobile thermal processing system that uses electrically-powered silicon carbide rods to heat organic wastes to combustion temperatures. Waste is fed into the primary chamber and exposed to infrared radiant heat (up to 1,850 °F) provided by silicon carbide rods above the conveyor belt. A blower delivers air to selected locations along the belt to control the oxidation rate of the waste feed.

Rotary Kilns

Commercial incinerator designs are rotary kilns, equipped with an afterburner, a quench, and an air pollution control system. The rotary kiln is a refractory-lined, slightly-inclined, rotating cylinder that serves as a combustion chamber and operates at temperatures up to 980 °F (1,800 °F). Incinerator off-gas requires treatment by an air pollution-control system to remove particulates and neutralize and remove acid gases (HCl, NOx, and SOx). Baghouses, venturi scrubbers, and wet electrostatic precipitators remove particulates; packed-bed scrubbers and spray driers remove acid gases.

Incineration, primarily off-site, has been selected or used as the remedial action at more than 150 Superfund sites. Incineration is subject to a series of technology-specific regulations, including the following federal requirements: CAA (air emissions), TSCA (PCB treatment and disposal), RCRA (hazardous waste generation, treatment, storage, and disposal), NPDES (discharge to surface waters), and NCA (noise). The duration of incineration technology ranges from short- to longterm Incineration is used to remediate soils contaminated with explosives and hazardous wastes, particularly chlorinated hydrocarbons, PCBs, and dioxins.

Factors that may limit the applicability and effectiveness of the process include:

- Only one off-site incinerator is permitted to burn PCBs and dioxins.
- There are specific feed size and materials handling requirements that can impact applicability or cost at specific sites.
- Heavy metals can produce a bottom ash that requires stabilisation.
- Volatile heavy metals, including lead, cadmium, mercury, and arsenic, leave the combustion unit with

the flue gases and require the installation of gas cleaning systems for removal.

- Metals can react with other elements in the feed stream, such as chlorine or sulfur, forming more volatile and toxic compounds than the original species. Such compounds are likely to be short- lived reaction intermediates that can be destroyed in a caustic quench.
- Sodium and potassium form low melting point ashes that can attack the brick lining and form a sticky particulate that fouls gas ducts

Permeable Reactive Barriers (PRB)

The extraction and treatment of contaminated ground water at hazardous waste sites continues to be an extremely costly endeavor, the effectiveness of which is often dubious at best. Permeable reaction walls are an emerging technology for the treatment of contaminated groundwater. Successful treatment of contaminated groundwater using this technique requires that the contaminant be rendered innocuous or immobile during transport through the in-situ treatment zone.

The extent of treatment, and the success of the permeable barrier system depends on the nature of the contaminant, the selection of the reactive material, the physical design of the treatment system, and natural site conditions. Instead of trying to remove the contaminated water from the subsurface for above-ground treatment, the emplacement of a permeable reactive geochemical barrier which intercepts the contaminated plume and transforms the contaminant to a nontoxic form is attractive for a number of reasons.

First, the in situ approach requires no above ground treatment facilities and the space can be returned to its original use. Second, there is no need for expensive above-ground treatment, storage, transport or disposal. Third, there is little or no operation and maintenance costs. A Permeable Reactive Barrier is an engineered treatment zone of reactive material(s) that is placed in the subsurface in order to remediate contaminated fluids as they flow through it.

A PRB has a negligible overall effect on bulk fluid flow rates in the subsurface strata, which is typically achieved by construction of a permeable reactive zone, or by construction of a permeable reactive 'cell' bounded by low permeability barriers that direct the contaminant towards the zone or reactive media" A PRB is built by digging a long, narrow trench in the path of the polluted groundwater.

Thetrench is filled with a reactive material that can clean up the harmful chemicals. Iron, lime-stone, and carbon are common types of reactive materials that can be used. The reactive materials may be mixed with sand to make it easier for water to flow through the wall, rather than around it. At some sites, the wall is part of a funnel that directs the polluted groundwater to the reactive part of the wall. The filled trench or funnel is covered with soil, so it usually cannot be seen above ground.

The material used to fill the trench depends on the types of harmful chemicals in the groundwater. Different materials clean up pollution through different methods by:

— Trapping or sorbing chemicals on their surface. For example, carbon has a surface that chemicals sorb to as groundwater passes through.

— Precipitating chemicals that are dissolved in water. This means the chemicals settle out of the groundwater as solid materials, which get trapped in the wall. For example, limestone can cause dissolved metals to precipitate.

— Changing the chemicals into harmless ones. For example, iron can change some types of solvents into harmless chemicals.

— Encouraging tiny bugs or microbes in the soil to eat the chemicals. For example, nutrients and oxygen in a PRB help the microbes grow and eat more chemicals. When microbes completely digest the chemicals, they can change them into water and harmless gases such as carbon dioxide.

The most common of the permeable barrier walls is the Iron Treatment Wall. It is made up of zero-valent iron or iron-bearing minerals that reduce chlorinated contaminants such as perchloroethylene (PCE). As the iron is oxidised, a chlorine atom is removed from the compound using electrons supplied by the oxidation of iron. The chlorinated compounds are reduced to nontoxic by-products.

Considerable research during the past several years has focused on the degradation of chlorinated solvents, such as TCE and PCE, by reactions at the surfaces of Fe(0). Although met with initial skepticism, the degradation process is now widely accepted as abiotic reductive dehalogenation, involving corrosion of the Fe(0) by the chlorinated hydrocarbon.

Iron corrosion processes in aqueous systems have been studied extensively. Until recently, the fate of corrosion processes in dilute aqueous concentrations of chlorinated solvents acting as the oxidising agents have not been investigated. The net reductive dechlorination reaction promoted by Fe(0) (Equation 3) may be viewed as the sum of anodic and cathodic reactions occurring at the iron metal surface (Equations 1 and 2, respectively), resulting in hydrocarbon products if the dechlorination proceeds to completion.

$$Fe^{0} \rightarrow Fe^{2+} + 2e \text{ - Anodic Reaction} \quad (1)$$

$$RCl + 2e^{-} + H^{+} \rightarrow RH + Cl \text{ - Cathodic Reaction} \quad (2)$$

$$Fe^{0} + RCl + H^{+} \rightarrow Fe^{2+} + RH + Cl \text{ - Net Reaction} \quad (3)$$

Under aerobic conditions, dissolved oxygen is usually the preferred electron acceptor and can compete with the chlorinated hydrocarbon as the favored oxidant (Equation 4). Indeed, chlorinated hydrocarbons such as PCE and carbon tetrachloride have oxidising potentials very similar to that of O_2. When sufficient oxygen is present, the Fe^{2+} generated in Equation 4 further oxidises to Fe^{3+} (Equation 5) and can precipitate as ferric hydroxide or (oxy)hydroxides (Equation 6) at the elevated pH typical of corroding Fe systems.

Corrosion of the iron can generate large amounts of iron oxides and (oxy)hydroxide precipitates that can exert significant additional chemical and physical effects within the reactive system. The rapid consumption of dissolved oxygen at the entrance to an iron system (column or barrier) has been shown to result in these precipitates that might impact a system's hydraulic performance at its upgradient interface.

$$2Fe^{0} + O_2 + 2H_2O \rightarrow 2Fe^{2+} + 4OH^{-} \quad (4)$$

$$4Fe^{2+} + 4H^{+} + O_2 \rightarrow 4Fe^{3+} + 2H_2O. \quad (5)$$

$$Fe^{3+} + 3OH \dashrightarrow Fe(OH)_{3(s)} \quad (6)$$

$$Fe^{0} + 2H_2O \rightarrow Fe^{2+} + H_2 + 2OH^{-} \quad (7)$$

$$Fe^{2+} + 2OH^{-} \rightarrow Fe(OH)_{2(s)} \quad (8)$$

Anaerobic corrosion of iron by water (Equation 7) proceeds slowly. Both reactions 4 and 7 result in an increased pH in weakly buffered systems, yielding ferric (oxy)hydroxides in aerobic systems (Equation 6) and ferrous (oxy)hydroxides in anaerobic systems (Equation8). The aqueous corrosion of iron is mediated by the layer of oxides, hydroxides and oxyhydroxides that are present at the iron-water interface.

The formation of these precipitates might further occlude the iron surface and affect its reduction-oxidation properties. However, this passive coating appears to be converted to magnetite, which is non-passivating, and seems to allow sufficient contaminant degradation rates that can be sustained over years of operation in the ground.

References

Crossley D.A., *Roles of Microflora and fauna in soil systems,* International Symposium on Pesticides in Soils, Feb. 25, 1970, University of Michigan.

Negraa, Christine; Donald S. Rossa and Antonio Lanzirottib, "Oxidizing Behavior of Soil Manganese: Interactions among Abundance, Oxidation State, and pH", *Soil Science Society of America Journal* 1 (69): 97-95. Retrieved on 2006-11-11, 2005.

Rainer Stegmann, *Treatment of Contaminated Soil: Fundamentals, Analysis, Applications,* Springer Verlag, Berlin 2001.

11

Soil Bioengineering

Soil bioengineering, in the context of upland slope protection and erosion reduction, combines mechanical, biological, and ecological concepts to arrest and prevent shallow slope failures and erosion. Basic approaches to upland slope protection and erosion control can be divided into two general categories: living and nonliving. Frequently, living and nonliving measures are combined to form a system. The living approach, which uses live plant materials, can be further divided into two specific categories: vegetative plantings and soil bioengineering.

Vegetative plantings are conventional plantings of grasses, forbs, and shrubs used to prevent surface erosion. Soil bioengineering utilises live plant parts to provide soil reinforcement and prevent surface erosion. In soil bioengineering systems, the installation may play the major structural roles immediately or may become the major structural component over time. Live staking, live fascines, brushlayers, branchpacking, and live gully repair are soil bioengineering techniques that use stems or branch parts of living plants as initial and primary soil reinforcing and stabilising material.

When these vegetative cuttings are placed in the ground, roots develop and foliage sprouts. The resulting vegetation becomes a major structural component of the soil bioengineering system. Live cribwalls, vegetated rock gabions, vegetated rock walls, and joint plantings are soil bioengineering techniques that use porous structures with openings through which vegetative cuttings are inserted and established. The inert structural elements provide immediate resistance to sliding, erosion, and washout.

As vegetation becomes established, roots invade and permeate the slope, binding it together into a unified, coherent mass. Over time, the structural elements diminish in importance as the vegetation increases in strength and functionality. Nonliving approaches use rigid constructions, such as surface armoring, gravity retaining walls, and rock buttresses. Vegetation can be used in conjunction with nonliving structures to create vegetated structures. Vegetation enhances the structures and helps reduce surface erosion, but usually does not provide any major reinforcement benefits.

Design Engineering Requirements

Soil bioengineering combines biological elements with engineering design principles. For example, engineering requirements may dictate highly compacted soil for fill slopes, while plants prefer relatively loose soil. Using a sheep's foot roller for compaction is a solution that would integrate biological and engineering requirements because it compacts the soil, but also allows plant establishment in resulting depressions in the slope. Differing needs can generally be integrated through creative approaches and occasional compromises in planning and design.

Applications

The soil bioengineering techniques in this document are generally appropriate for immediate protection of slopes against surface erosion, shallow mass wasting, cut and fill slope stabilisation, earth embankment protection, and small gully repair treatment. Other situations where soil bioengineering measures can be employed are not discussed in this chapter. These situations include dune stabilisation, wetland buffers, reservoir drawdown areas where plants can be submerged for extended periods, and areas with highly toxic soils. Properly designed and constructed soil bioengineering measures have also been employed with considerable success in stabilising shorelines and streambanks.

Soil Bioengineering Systems

Soil bioengineering uses particular characteristics of vegetative components and integrates specific characteristics of structures with vegetation. The resulting systems and their components have benefits and limitations that need to be considered prior to selecting them for use.

Vegetative Components

Herbaceous Species

Herbaceous vegetation, especially grasses and forbs, offers long-term protection against surface (water and wind) erosion on slopes. It provides only minor protection against shallow mass movement. Vegetation helps to prevent surface erosion by:

— Binding and restraining soil particles in place

— Reducing sediment transport

— Intercepting raindrops

— Retarding velocity of runoff

— Enhancing and maintaining infiltration capacity

Minimising freeze-thaw cycles of soils susceptible to frost Herbaceous species are almost always used in conjunction with soil bioengineering projects to add protection against surface erosion.

Woody Species

More deeply rooted woody vegetation provides greater protection against shallow mass movement by:

— Mechanically reinforcing the soil with roots

— Depleting soil-water through transpiration and interception

— Buttressing and soil arching action from embedded stems

Live fascines, for example, provide many of these protective functions. They are fabricated from woody species, such as shrub willow or shrub dogwood, into sausage-like bundles, which are placed with the stems oriented generally parallel to the slope contour. This method of placement and orientation would not be used in slope reinforcement. Live fascines serve to dissipate the energy of downward moving water by trapping debris and providing a series of benches on which grasses, seedlings, and transplants establish more easily.

Portions of the live fascines also root and become part of the stabilising cover. Live fascines provide an immediate increase in surface stability and can further improve soil stability to depths of 2 to 3 feet as roots develop. In the case of brushlayering, live branches or shoots of such woody species as shrub willow, dogwood, or privet are placed in successive layers with the stems generally oriented perpendicular to the slope contour. This orientation is the optimal direction for maximum reinforcing effect in a slope. Brushlayering can improve soil stability to depths of 4 to 5 feet.

Structural Components

Properly designed and installed structures help to stabilise a slope against shallow mass movement and protect the slope against rill and gully formation. Structures also play a critical role in the establishment of vegetation on steep slopes or in areas subject to severe erosion. They may make it possible to establish plants on slopes steeper than would normally be possible. Structures stabilise slopes during the critical time for seed germination and root growth. Without this stabilisation, vegetative plantings would fail during their most vulnerable time.

Materials

Structures can be built from natural or manufactured materials. Natural materials, such as earth, rock, stone, and timber, usually cost less, are environmentally more compatible, and are better suited to vegetative treatment or slight modifications than are manufactured materials. Natural materials may also be available onsite at no cost. Some structures are comprised of both natural and manufactured materials. Examples include concrete cribwalls, steel bin walls, gabion walls or revetments, welded wire or polymeric geogrid walls, and reinforced earth. In these cases steel and concrete mostly provide rigidity, strength, and reinforcement, whereas stone, rock, and soil provide mass. These types of structures have spaces that are often planted with herbaceous or woody vegetation.

Retaining Structures

A retaining structure of some type is usually required to protect and stabilise steep slopes. Low retaining structures at the toe of a slope make it possible to grade the slope back to a more stable angle that can be successfully revegetated without loss of land at the crest. Structures are generally capable of resisting much higher lateral earth pressures and shear stresses than vegetation.

Grade Stabilisation Structures

Grade stabilisation structures are used to control and prevent gully erosion. A grade stabilisation structure reduces the grade above it and dissipates the excess energy of flowing water within the structure itself. Debris and sediment tend to be deposited and trapped upstream of the structure. This, in turn, permits establishment of vegetation behind the structure, which further stabilises the ground.

Grade stabilisation structures may range from a series of simple timber check dams to complex concrete overfall structures and earth embankments with pipe spillways. Gully control provides a good example of the integration of structures and vegetation. Structural measures may be required in the short term to stabilise critical locations. The long-term goal is to establish and maintain a vegetative cover that prevents further erosion. This goal is seldom realised unless the severe gully conditions can be altered immediately. Vegetation alone, for example, will rarely stabilise gully headcuts because of the concentrated water flow, overfalls, and pervasive forces that promote gully enlargement in an unstable channel system. Initially, the vegetation and the structure work together in an integrated fashion. The ultimate function of these structures, however, is to help establish vegetation which will provide longterm protection.

Attributes and Limitations

Soil bioengineering measures should not be viewed as a panacea or solution for all slope failure and surface erosion problems. Soil bioengineering has unique attributes, but is not appropriate for all sites and situations. In certain cases, a conventional vegetative treatment works satisfactorily at less cost. In other cases, the more appropriate and most effective solution is a structural retaining system alone or in combination with soil bioengineering. The following specific attributes and limitations should be considered before applying a soil bioengineering technique:

Environmental Compatibility

Soil bioengineering systems generally require minimal access for equipment and workers and cause relatively minor site disturbance during installation. These are generally priority considerations in environmentally sensitive areas, such as parks, woodlands, riparian areas, and scenic corridors where aesthetic quality, wildlife habitat, and similar values may be critical.

Cost Effectiveness

Field studies have shown instances where combined slope protection systems have proven more cost effective than the use of either comparative vegetative treatments or structural solutions alone. Where construction methods are labour intensive and labour costs are reasonable, the combined systems may be especially cost effective. Where labour is either scarce or extremely expensive, however, soil bioengineering systems may be less practical than structural measures. This can be offset by the time of year when other construction work is slow. Using indigenous materials accounts for some of the cost effectiveness because plant costs are limited to labour for harvesting, handling, and direct costs for transporting the plants to the site.

Planting Times

Soil bioengineering systems are most effective when they are installed during the dormant season, usually the late fall, winter, and early spring. Constraints on planting times or the availability of the required quantities of suitable plant materials during allowable planting times may limit the usefulness of soil bioengineering methods.

Difficult Sites

Soil bioengineering is often a useful alternative for small, highly sensitive, or steep sites where the use of machinery is not feasible and hand labour is a necessity. However, rapid vegetative establishment may be difficult on extremely steep slopes. The usefulness of soil bioengineering methods may be limited by the available medium for plant growth, such as rocky or gravelly slopes that lack sufficient fines or moisture to support the required plant growth. In addition, soil-restrictive layers, such as hardpans, may prevent required root growth. The biotechnical usefulness of vegetation would be limited on slopes that are exposed to high velocity water flow or constant inundation.

Harvesting Local Plant Material

Appropriate vegetation is often obtained from local stands of willows and other suitable species. This stock is already well suited to the climate, soil conditions, and available moisture and is a good candidate for survival. While harvesting may often help a beneficial species proliferate, reliance on the use of local plant materials and gathering in the wild could result in short supplies or unacceptable depletion of site vegetation. Some localities may have prohibitions against gathering native plants, and materials must be purchased from commercial sources.

Biotechnical Strengths

Soil bioengineering systems are strong initially and grow stronger with time as vegetation becomes established. In some instances, the primary role of the structural component is to give the vegetation a better chance to become established. It has been shown in slope reconstruction projects that soil bioengineering systems can withstand heavy rainfalls immediately after installation. Even if established vegetation dies, the plant roots and surface residue may continue to play an important protective role during reestablishment.

Maintenance Requirements

Once vegetation is well established on a soil bioengineering project, usually within one growing season, it generally becomes self-repairing by regeneration and growth and requires little maintenance. However, a newly installed soil bioengineering project will require careful periodic inspections until it is established. Established vegetation is vulnerable to trampling, drought, grazing, nutrient deficiencies, toxins, and pests, and may require special management measures at times.

Basic Principles of Soil Erosion Control

The basic principles that apply to conventional soil erosion control also apply in general to soil bioengineering. These principles are mostly common sense guidelines that involve planning, timing, and minimising site disturbance as well as the design of individual measures themselves. Applicable principles can be summarised as follows:

Fit the Soil Bioengineering System

This means considering site topography, geology, soils, vegetation, and hydrology. Avoid extensive grading and earthwork in critical areas and perform soil tests to determine if vigorous plant growth can be supported. At a minimum, collect the following information:

Topography and Exposure

— Note the degree of slope in stable and unstable areas. Also note the presence or lack of moisture. The likely success of soil bioengineering treatments can best be determined by observing existing stable slopes in the vicinity of the project site.

— Note the type and density of existing vegetation in areas with and without moisture and on slopes facing different directions. Certain plants grow well on east-facing slopes, but will not survive on south-facing slopes.

— Look for areas of vegetation that may be growing more vigorously than other site vegetation. This is generally a good indicator of excess moisture, such as seeps and a perched watertable, or it may reflect a change in soils.

Geology and Soils

— Consult SCS geologists about geologic history and types of deposits (colluvium, glacial, alluvium, other).

— Note evidence of past sliding. If site evidence exists, determine whether the slide occurred along a deep or shallow failure surface. Leaning or deformed trees may indicate previous slope movement or downhill creep. In addition to site evidence, check aerial photos, which can reveal features that may not be apparent from a site visit.

— Determine soil type and depth. Use the soil survey report, if available, or consult SCS soil scientists.

Hydrology

— Determine the drainage area associated with the problem area. Note whether water can be diverted away from the problem area.

— Determine the annual precipitation. Are there concentrated discharges?
— Calculate peak flows or mean discharge through the project area.
— If a seep area was noted, locate the source of the water. Determine whether the water can be intercepted and diverted away from the slope face.

Vegetation

Vegetation provides excellent protection against surface erosion and shallow slope failures. Soil bioengineering measures are designed to aid or enhance the reestablishment of vegetation.

Limit Removal

— Limit cleared area to the smallest practical size
— Limit duration of disturbance to the shortest practical time
— Remove and store existing woody vegetation that may be used later in the project
— Schedule land clearing during periods of low precipitation whenever possible

Protect Topsoil

Topsoil removed during clearing and grading operations can be reused during planting operations.

Protect Areas during Construction

Temporary erosion and sediment control measures can be used.

Divert, Drain, or Store Excess Water

— Install a suitable system to handle increased and/or concentrated runoff caused by changed soil and surface conditions during and after construction.

— Install permanent erosion and sediment control measures in the project before construction is started if possible.

Design Considerations

Earthwork

Typically, sites require some earthwork prior to the installation of soil bioengineering systems. A steep undercut or slumping bank, for example, requires grading to flatten the slope for stability. The degree of flattening depends on the soil type, hydrologic conditions, geology, and other site factors.

Scheduling and Timing

Planning and coordination are needed to achieve optimal timing and scheduling. The seasonal availability of plants or the best time of year to install them may not coincide with the construction season or with tight construction schedules. In some cases, rooted stock may be used as an alternative to unrooted dormant season cuttings.

Vegetative Damage

Vegetative damage to inert structures may occur when inappropriate species or plant materials that exceed the size of openings in the face of structures are used. Vegetative damage does not generally occur from roots. Plant roots tend to avoid porous, open-faced retaining structures because of excessive sunlight, moisture deficiencies, and the lack of a growing medium.

Moisture Requirements

The backfill behind a stable retaining structure has certain specified mechanical and hydraulic properties. Ideally, the fill is coarse-grained, free-draining, granular material. Excessive amounts of clay, silt, and organic matter are not desirable. Free drainage is essential to the mechanical integrity of an earthretaining structure and also important to vegetation, which cannot tolerate waterlogged soil conditions.

The establishment and maintenance of vegetation, however, usually requires the presence of some fines and organic matter in the soil to provide adequate moisture and nutrient retention. In many instances, these biological requirements can be satisfied without compromising engineering performance of the structure.

With cribwalls, for example, adequate amounts of fines or other amendments can be incorporated into the backfill. Gabions can be filled with rock and soil drifted into them to facilitate growth of vegetation. Woody vegetative cuttings can be placed between the baskets during filling and into the soil or backfill beyond the baskets. The needs of plants and the requirements of structures must be taken into account when designing a system.

Techniques and Materials

Soil bioengineering measures have certain requirements and capabilities. Plant species must be suitable for the intended use and adapted to the site's climate and soil conditions. Consult a plant materials specialist about available cultivars to ensure that appropriate species are used. Species that root easily, such as willow, are required for such measures as live fascines, brushlayer, and live staking or where unrooted stems are used with structural measures.

Rooted plants and vegetative cuttings are living materials and must be handled properly to avoid excess stress, such as drying or exposure to heat. They must be installed in moist soil and be adequately covered. The soil must be compacted to eliminate or minimise air pockets around the buried stems. If soils are not at or near moisture capacity, the installation must be delayed unless deep and regular irrigation can be provided during and following installation. Installation of soil bioengineering systems is best accomplished in the late fall at the onset of plant dormancy, in the winter as long as the ground is not frozen, or in early spring before growth begins. In some cases installation after initial spring growth may be successful if extreme care is used, but the risks of failure are high. Summer installation is not recommended. Rooted plants can be used, but they are sometimes less effective and more expensive. All installations should be inspected regularly and provisions made for prompt repair if needed.

Initial failure of a small portion of a system normally can be repaired easily and inexpensively. Neglect of small failures, however, can result in the failure of large portions of a system. Properly designed and installed vegetative portions of systems will become self-repairing to a large extent. Periodic pruning and replanting may be required to maintain healthy and vigorous vegetation. Structural elements, such as cribwalls, rock walls, and gabions, may require maintenance and/or replacement throughout their life. Where the main function of structural elements is to allow vegetation to become established and take over the role of slope stabilisation, the eventual deterioration of the structures is not a cause for concern.

Soil Bioengineering Techniques

The following describes soil bioengineering techniques. Their applications, effectiveness, and construction guidelines are also presented.

Live Stake

Live staking involves the insertion and .tamping of live, rootable vegetative cuttings into the ground. If correctly prepared and placed, the live stake will root and grow. A system of stakes creates a living root mat that stabilises the soil by reinforcing and binding soil particles together and by extracting excess soil moisture. Most willow species root rapidly and begin to dry out a slope soon after installation. This is an appropriate technique for repair of small earth slips and slumps that frequently are wet.

Applications and Effectiveness:

— A technique for relatively uncomplicated site conditions when construction time is limited and an inexpensive method is necessary.

— May be used for pegging down surface erosion control materials.
— Enhances conditions for natural invasion and the establishment of other plants from the surrounding plant community.
— Can be used to stabilise intervening area between other soil bioengineering techniques, such as live fascines.

Construction Guidelines

Live material sizes: The cuttings are usually ½ to 1 ½ inches in diameter and 2 to 3 feet long. For final size determination, refer to the available cutting source.

Live Material Preparation

— The materials must have side branches cleanly removed and the bark intact.
— The basal ends should be cut at an angle for easy insertion into the soil. The top should be cut square.
— Materials should be installed the same day that they are prepared.

Installation

— Tamp the live stake into the ground at right angles to the slope. The installation may be started at any point on the slope face.
— The live stakes should be installed 2 to 3 feet apart using triangular spacing. The density of the installation will range from 2 to 4 stakes per square yard.
— The buds should be oriented up.
— Four-fifths of the length of the live stake should be installed into the ground and soil firmly packed around it after installation.
— Do not split the stakes during installation.
— Stakes that split should be removed and replaced.
— An iron bar can be used to make a pilot hole in firm soil. Drive the stake into the ground with a dead blow hammer.

Live Fascine

Live fascines are long bundles of branch cuttings bound together into sausage-like structures. When cut from appropriate species and properly installed with live and dead stout stakes, they will root and immediately begin to stabilise slopes. They should be placed in shallow contour trenches on dry slopes and at an angle on wet slopes to reduce erosion and shallow face sliding. This system, installed by a trained crew, does not cause much site disturbance.

Applications and Effectiveness

- An effective stabilisation technique for slopes.
- Protects slopes from shallow slides (1 to 2 foot depth).
- Immediately reduces surface erosion or rilling.
- Suited to steep, rocky slopes, where digging is difficult.
- Capable of trapping and holding soil on the face of the slope, thus reducing a long slope into a series of shorter slopes.
- Enhances vegetative establishment by creating a microclimate conducive to plant growth.

Construction Guidelines

Live materials-Cuttings must be from species, such as young willows or shrub dogwoods, that root easily and have long, straight branches.

Live Material Sizes and Preparation

- Cuttings tied together to form live fascine bundles vary in length from 5 to 30 feet or longer, depending on site conditions and limitations in handling.
- The completed bundles should be 6 to 8 inches in diameter, with all of the growing tips oriented in the same direction. Stagger the cuttings in the bundles so that tops are evenly distributed throughout the length of the uniformly sized live fascine.
- Live stakes should be 2 ½ feet long in cut slopes and 3 feet long in fill slopes.

- Inert materials-String used for bundling should be untreated twine.
- Dead stout stakes used to secure the live fascines should be 2 ½-foot long, untreated, 2 by 4 lumber. Each length should be cut again diagonally across the 4-inch face to make two stakes from each length. Only new, sound, unused lumber should be used, and any stakes that shatter upon installation should be discarded.

Installation

- Prepare the live fascine bundles and live stakes immediately before installation.
- Beginning at the base of the slope, dig a trench on the contour just large enough to contain the live fascine. The trench will vary in width from 12 to 18 inches, depending on the angle of the slope to be treated. The depth will be 6 to 8 inches, depending on the individual bundle's final size.
- Place the live fascine into the trench.
- Drive the dead stout stakes directly through the live fascine every 2 to 3 feet along its length. Extra stakes should be used at connections or bundle overlaps. Leave the top of the stakes flush with the installed bundle.
- Live stakes are generally installed on the downslope side of the bundle. Drive the live stakes below and against the bundle between the previously installed dead stout stakes. The live stakes should protrude 2 to 3 inches above the top of the live fascine. Place moist soil along the sides of the live fascine. The top of the fascine should be slightly visible when the installation is completed.

Next, at intervals on contour or at an angle up the face of the bank, repeat the preceding steps to the top of the slope. When possible, place one or two rows over the top of the slope. Long straw or similar mulching material should be placed between rows on 2.5:1 or flatter slopes, while slopes steeper than 2.5:1

should have jute mesh or similar material placed in addition to the mulch.

Brushlayering

Brushlayering is somewhat similar to live fascine systems because both involve the cutting and placement of live branch cuttings on slopes. The two techniques differ principally in the orientation of the branches and the depth to which they are placed in the slope. In brushlayering, the cuttings are oriented more or less perpendicular to the slope contour. The perpendicular orientation is more effective from the point of view of earth reinforcement and mass stability of the slope.

Brushlayering consists of placing live branch cuttings in small benches excavated into the slope. The benches can range from 2 to 3 feet wide. These systems are recommended on slopes up to 2:1 in steepness and not to exceed 15 feet in vertical height. Brushlayer branches serve as tensile inclusions or reinforcing units. The portions of the brush that protrude from the slope face assist in retarding runoff and reducing surface erosion.

Applications and effectiveness-Brushlayers perform several immediate functions in erosion control, earth reinforcement, and mass stability of slopes:

— Breaking up the slope length into a series of shorter slopes separated by rows of brushlayer.

— Reinforcing the soil with the unrooted branch stems.

— Reinforcing the soil as roots develop, adding significant resistance to sliding or shear displacement.

— Providing slope stability and allowing vegetative cover to become established.

— Trapping debris on the slope.

— Aiding infiltration on dry sites.

— Drying excessively wet sites.

— Adjusting the site's microclimate, thus aiding seed germination and natural regeneration.

— Redirecting and mitigating adverse slope seepage by acting as horizontal drains.

Construction Guidelines

Live material sizes: Branch cuttings should be ½ to 2 inches in diameter and long enough to reach the back of the bench. Side branches should remain intact for installation.

Installation

— Starting at the toe of the slope, benches should be excavated horizontally, on the contour, or angled slightly down the slope, if needed to aid drainage. The bench should be constructed 2 to 3 feet wide.

— The surface of the bench should be sloped so that the outside edge is higher than the inside.

— Live branch cuttings should be placed on the bench in a crisscross or overlapping configuration.

— Branch growing tips should be aligned toward the outside of the bench.

— Backfill is placed on top of the branches and compacted to eliminate air spaces. The brush tips should extend slightly beyond the fill to filter sediment.

— Each lower bench is backfilled with the soil obtained from excavating the bench above.

— Long straw or similar mulching material with seeding should be placed between rows on 3:1 or flatter slopes, while slopes steeper than 3:1 should have jute mesh or similar material placed in addition to the mulch.

— The brushlayer rows should vary from 3 to 5 feet apart, depending upon the slope angle and stability.

Branchpacking

Description-Branchpacking consists of alternating layers of live branch cuttings and compacted backfill to repair small localised slumps and holes in slopes.

Applications and Effectiveness

- — Effective in earth reinforcement and mass stability of small earthen fill sites.
- — Produces a filter barrier, reducing erosion and scouring conditions.
- — Repairs holes in earthen embankments other than dams where water retention is a function.
- — Provides immediate soil reinforcement.

Construction Guidelines

Live material-Live branch cuttings may range from ½ inch to 2 inches in diameter. They should be long enough to touch the undisturbed soil at the back of the trench and extend slightly from the rebuilt slope face. Inert material-Wooden stakes should be 5 to 8 feet long and made from 3- to 4-inch diameter poles or 2 by 4 lumber, depending upon the depth of the particular slump or hole.

Installation

- — Starting at the lowest point, drive the wooden stakes vertically 3 to 4 feet into the ground. Set them 1 to 1 ½ feet apart.
- — A layer of living branches 4 to 6 inches thick is placed in the bottom of the hole, between the vertical stakes, and perpendicular to the slope face. They should be placed in a crisscross configuration with the growing tips generally oriented toward the slope face. Some of the basal ends of the branches should touch the back of the hole or slope.
- — Subsequent layers of branches are installed with the basal ends lower than the growing tips of the branches.
- — Each layer of branches must be followed by a layer of compacted soil to ensure soil contact with the branch cuttings.
- — The final installation should match the existing slope. Branches should protrude only slightly from the filled face.

— The soil should be moist or moistened to insure that live branches do not dry out.

The live branch cuttings serve as tensile inclusions for reinforcement once installed. As plant tops begin to grow, the branchpacking system becomes increasingly effective in retarding runoff and reducing surface erosion. Trapped sediment refills the localised slumps or holes, while roots spread throughout the backfill and surrounding earth to form a unified mass. Branchpacking is not effective in slump areas greater than 4 feet deep or 5 feet wide.

Live Gully Repair

A live gully repair utilises alternating layers of live branch cuttings and compacted soil to repair small rills and gullies. Similar to branchpacking, this method is more appropriate for the repair of rills and gullies.

Applications and Effectiveness

— The installed branches offer immediate reinforcement to the compacted soil and reduce the velocity of concentrated flow of water.

— Provides a filter barrier that reduces rill and gully erosion.

— Limited to rills or gullies which are a maximum of 2 feet wide, 1 foot deep, and 15 feet long.

Construction Guidelines

Live material sizes-Live branch cuttings may range from ½ inch to 2 inches in diameter. They should be long enough to touch the undisturbed soil at the back of the rill or gully and extend slightly from the rebuilt slope face. Inert materials-Fill soil is compacted in alternate layers with live branch cuttings.

Installation

— Starting at the lowest point of the slope, place a 3- to 4-inch layer of branches at lowest end of the rill or gully and perpendicular to the slope.

— Cover with a 6- to 8-inch layer of fill soil.
— Install the live branches in a crisscross fashion.
— Orient the growing tips toward the slope face with basal ends lower than the growing tips.
— Follow each layer of branches with a layer of compacted soil to ensure soil contact with the live branch cuttings.

Live Cribwall

A live cribwall consists of a hollow, box-like interlocking arrangement of untreated log or timber members. The structure is filled with suitable backfill material and layers of live branch cuttings which root inside the crib structure and extend into the slope. Once the live cuttings root and become established, the subsequent vegetation gradually takes over the structural functions of the wood members.

Applications and Effectiveness

— This technique is appropriate at the base of a slope where a low wall may be required to stabilise the toe of the slope and reduce its steepness.
— Not designed for or intended to resist large, lateral earth stresses. It should be constructed to a maximum of 6 feet in overall height, including the excavation required for a stable foundation.
— Useful where space is limited and a more vertical structure is required.
— Provides immediate protection from erosion, while established vegetation provides longterm stability.
— Should be tilted back or battered if the system is built on a smooth, evenly sloped surface.
— May also be constructed in a stair-step fashion, with each successive course of timbers set back 6 to 9 inches toward the slope face from the previously installed course.

Construction Guidelines

— Live material sizes-Live branch cuttings should be ½ to 2 inches in diameter and long enough to reach the back of the wooden crib structure.

— Inert materials-Logs or timbers should range from 4 to 6 inches in diameter or dimension. The lengths will vary with the size of the crib structure.

— Large nails or rebar are required to secure the logs or timbers together.

Installation

— Starting at the lowest point of the slope, excavate loose material 2 to 3 feet below the ground elevation until a stable foundation is reached.

— Excavate the back of the stable foundation (closest to the slope) slightly deeper than the front to add stability to the structure.

— Place the first course of logs or timbers at the front and back of the excavated foundation, approximately 4 to 5 feet apart and parallel to the slope contour.

— Place the next course of logs or timbers at right angles (perpendicular to the slope) on top of the previous course to overhang the front and back of the previous course by 3 to 6 inches.

— Each course of the live cribwall is placed in the same manner and nailed to the preceding course with nails or reinforcement bars.

— When the cribwall structure reaches the existing ground elevation, place live branch cuttings on the backfill perpendicular to the slope; then cover the cuttings with backfill and compact.

Live branch cuttings should be placed at each course to the top of the cribwall structure with growing tips oriented toward the slope face. Follow each layer of branches with a layer of compacted soil to ensure soil contact with the live branch cuttings. Some of the basal ends of the live branch cuttings should reach to undisturbed soil at the back of the cribwall with growing tips protruding sightly beyond the front of the cribwall.

Vegetated Rock Gabions

Vegetated gabions begin as rectangular containers fabricated from a triple twisted, hexagonal mesh of heavily galvanised steel wire. Empty gabions are placed in position, wired to adjoining gabions, filled with stones and then folded shut and wired at the ends and sides. Live branches are placed on each consecutive layer between the rockfilled baskets. These will take root inside the gabion baskets and in the soil behind the structures. In time the roots consolidate the structure and bind it to the slope.

Applications and Effectiveness

— This technique is appropriate at the base of a slope where a low wall may be required to stabilise the toe of the slope and reduce its steepness.

— Not designed for or intended to resist large, lateral earth stresses. It should be constructed to a maximum of 5 feet in overall height, including the excavation required for a stable foundation.

— Useful where space is limited and a more vertical structure is required.

Construction Guidelines

— Live material sizes-Branches should range from ½ to 1 inch in diameter and must be long enough to reach beyond the back of the rock basket structure into the backfill.

— Inert materials-Inert material requirements include wire gabion baskets and rocks to fill the baskets.

Installation

— Starting at the lowest point of the slope, excavate loose

material 2 to 3 feet below the ground elevation until a stable foundation is reached.

— Excavate the back of the stable foundation (closest to the slope) slightly deeper than the front to add stability to the structure. This will provide additional stability to the structure and ensure that the living branches root well.
— Place the fabricated wire baskets in the bottom of the excavation and fill with rock.
— Place backfill between and behind the wire baskets.
— Place live branch cuttings on the wire baskets perpendicular to the slope with the growing tips oriented away from the slope and extending slightly beyond the gabions. The live cuttings must extend beyond the backs of the wire baskets into the fill material. Place soil over the cuttings and compact it.
— Repeat the construction sequence until the structure reaches the required height.

Vegetated Rock Wall

A vegetated rock wall is a combination of rock and live branch cuttings used to stabilise and protect the toe of steep slopes. Vegetated rock walls differ from conventional retaining structures in that they are placed against relatively undisturbed earth and are not intended to resist large lateral earth pressures.

Applications and Effectiveness

— This system is appropriate at the base of a slope where a low wall may be required to stabilise the toe of the slope and reduce its steepness.
— Useful where space is limited and natural rock is available.

Construction Guidelines

— Live material sizes-Live cuttings should have a diameter of ½ to 1 inch and be long enough to reach beyond the rock structure into the fill or undisturbed soil behind.

— Inert materials-Inert materials consist of rocks and fill material for the wall construction. Rock used should normally range from 8 to 24 inches in diameter. Larger boulders should be used for the base.

Installation

— Starting at the lowest point of the slope, remove loose soil until a stable base is reached. This usually occurs 2 to 3 feet below ground elevation. Excavate the back of the stable foundation (closest to the slope) slightly deeper than the front to add stability to the structure.

— Excavate the minimum amount from the existing slope to provide a suitable recess for the wall.

— Provide a well-drained base in locations subject to deep frost penetration.

— Place rocks with at least a three-point bearing on the foundation material or underlying rock course. They should also be placed so that their centre of gravity is as low as possible, with their long axis slanting inward toward the slope if possible.

— When a rock wall is constructed adjacent to an impervious surface, place a drainage system at the back of the foundation and outside toe of the wall to provide an appropriate drainage outlet.

— Overall height of the rock wall, including the footing, should not exceed 5 feet.

— A wall can be constructed with a sloping bench behind it to provide a base on which live branch cuttings can be placed during construction. Live branch cuttings should also be tamped or placed into the openings of the rock wall during or after construction. The butt ends of the branches should extend into the backfill or undisturbed soil behind the wall.

— The live branch cuttings should be oriented perpendicular to the slope contour with growing tips protruding slightly from the finished rock wall face.

Joint Planting

Joint planting or vegetated riprap involves tamping live cuttings of rootable plant material into soil between the joints or open spaces in rocks that have previously been placed on a slope. Alternatively, the cuttings can be tamped into place at the same time that rock is being placed on the slope face.

Applications and Effectiveness

Used where rock riprap is required. Roots improve drainage by removing soil moisture. Over time, they create a living root mat in the soil base upon which the rock has been placed. The root systems of this mat help to bind or reinforce the soil and to prevent washout of fines between and below the rock units.

Construction Guidelines

Live material sizes-The cuttings must have side branches removed and bark intact. They should range in diameter from ½ inch to 1 ½ inches and be sufficiently long to extend into soil below the rock surface.

Installation

- — Tamp live branch cuttings into the openings of the rock during or after construction. The butt ends of the branches should extend into the backfill or undisturbed soil behind the riprap.
- — Orient the live branch cuttings perpendicular to the slope with growing tips protruding slightly from the finished face of the rock.

Soil Bioengineering Materials

Locating and Selecting Plant Materials

Commercial sources-Commercially grown plant materials are suitable sources of vegetation for use in soil bioengineering systems; however, it is necessary to allow adequate lead time for their procurement and delivery. The SCS Plant Materials Programme has selected superior cultivars of willows, dogwoods, and other species, which have been evaluated in soil bioengineering systems and are being produced commercially. The most desirable species and cultivars to use can be determined from specifications for critical area stabilisation for each state.

Plant materials specialists are closely involved with the testing of plants and can assist with up-to-date information on cultivar adaptation. Harvesting indigenous species-Correctly selected indigenous species harvested from existing stands of living woody vegetation are the preferred soil bioengineering materials. The use of indigenous live materials requires careful selection, harvesting, handling, and transporting. They should result in plants that have deep and strong root systems, are relatively inexpensive, are usually effective, and can be installed quickly.

Live plant materials can be cut from existing native or naturalised stands found near the project site or within practical hauling distance. The source site must contain plant species that will propagate easily from cuttings. Cuttings are normally ½ to 2 inches in diameter and range in length from 2 to 6 feet. Chain saws, bush axes, loppers, and pruners are recommended for cutting living plant material. Safety precautions must be followed when using these tools. Onsite plant material should be harvested with great care. In some places a large area can be cut, but other sites require selective cutting.

Cuts should be made at a blunt angle, 8 to 10 inches from the ground, to assure that the source sites will regenerate rapidly and in a healthy manner. The harvesting site should be left clean and tidy. Remnant materials that are too large for use in soil bioengineering projects should be chipped or left in piles for wildlife cover. A site may be needed again for future harvesting and should be left in a condition that will enhance its potential for regeneration.

Binding and storage-Live cuttings should be bundled together securely at the collection site for easy loading and handling and for protection during transport. Side branches and brushy limbs should be kept intact.

Transporting-The bundles of live cuttings should be placed on the transport vehicles in an orderly fashion to prevent damage and facilitate handling. They should be covered with a tarpaulin during transportation to prevent drying and additional stress.

Handling-Live cuttings should arrive on the job site within 8 hours of harvest and should be installed immediately. This is especially critical when the ambient temperature is 50 °F or above.

Live cuttings not installed on the day they arrive should be promptly placed in controlled storage conditions and protected until they can be installed. When in storage, the cuttings must receive continuous shade, must be sheltered from the wind, and must be continuously protected from drying by being heeled into moist soils or stored in uncontaminated water. All live cuttings should be removed from storage and used within 2 days of harvest.

Installing Plant Materials

— Timing-Installation of live cuttings should begin concurrently with earth moving operations if they are carried out during the dormant season. All construction operations should be phased together whenever possible. The best time for installation of soil bioengineering systems is during the dormant season, which generally occurs from September to March throughout most of the United States. Each geographic area has a specific dormant season within this broad range, and yearly variations should be taken into account.

— Planting medium-Soil bioengineering projects ideally use onsite stockpiled topsoil as the planting medium of choice. Gravel is not a suitable material for use as fill around live plant materials. Soil bioengineering systems need to be installed in a planting medium that includes fines and organic material and is capable of supporting plant growth. Muddy soils that are otherwise suitable should not be used until they have been dried to a workable moisture content. Heavy clays should be mixed with organic soils to increase porosity. Select soil backfill does not need to be organic topsoil, but it must be able to support plant growth.

Soil samples of the onsite materials should be taken prior to installation of live woody cuttings. Soil samples should also be taken of all fill materials that are brought to the site prior to use. Nutrient testing by an approved laboratory should include analyses for a full range of nutrients, metal contents, and pH. The laboratory reports should also include recommended fertiliser and lime amendments for woody plant materials.

All fill soil around the live vegetative cuttings should be compacted to densities approximating the surrounding natural soil densities. The soil around plants should be free of voids.

Quality Control

Maintaining quality control throughout installation and maintenance operations will ensure a successful soil bioengineering project. The following guidelines are recommended:

Pre-construction

- Select plant species for conformance to requirements.
- Locate and secure source sites for harvesting live cuttings or commercial procurement.
- Define construction work area limits.
- Fence off sites requiring special protection.
- Complete and inspect the following preparations:
  - Layout
  - Excavation, systems excavation

- — Bench size, shape, angle
- — Preparation of site; i.e., clearing, grading, and shaping
- — Disposal of excess gravel, soil, and debris
- — Depth of excavation
- — Vegetation to be removed/preserved
- — Stockpiling of suitable soil and/or rock

Construction

Inspect each system component, at every stage, for the following:

- — Angle of placement and orientation of the live cuttings
- — Backfill material/rock and stone material
- — Fertiliser, method and quantity applied
- — Lime, method and quantity applied

Preparation of trenches or benches in cut and fill slopes

- — Staking
- — Pruning
- — Stock handling and preparation
- — Soil compaction
- — Watering
- — Ensure that proper maintenance occurs during and after installation.
- — Inspect daily for quality control.
    - — Check all cuttings; remove unacceptable material and use fresh stock for replacement installations.
    - — Continuously check all items in the preconstruction and construction inspection lists.
    - — Inspect the plant materials storage area when it is in use.

Establishment Period

Interim inspections-Inspections should be made after the soil bioengineering measures have been installed. The following schedule is recommended:

- Inspect biweekly for the first 2 months. Inspections should note insect infestations, soil moisture, and other conditions that could lead to poor survivability. Immediate action, such as the application of supplemental water, should be taken if conditions warrant.
- Inspect monthly for the next 6 months. Systems not in acceptable growing condition should be noted and, as soon as seasonal conditions permit, should be removed from the site and replaced with materials of the same species and sizes as originally specified.
- Needed reestablishment work should be performed every 6 months during the initial 2-year establishment period. This will usually consist of replacing dead material.
- Extra inspections should always be made during periods of drought or heavy rains. Damaged sections should always be repaired immediately.

Final inspection-A final inspection should be held 2 years after installation is completed. Healthy growing conditions should exist.

- Healthy growing conditions in all areas refer to overall leaf development and rooted stems defined as follows:

  Live stakes ------------------- 70%-100% growing
  Live fascines ---------------- 20%-50% growing
  Live cribwall ---------------- 30%-60% growing
  Brushlayers ----------------- 40%-70% growing
  Branchpacking ------------ 40%-70% growing
  Live gully repair ----------- 30%-50% growing
  Vegetated rock wall ------ 50%-80% growing
  Vegetated gabion --------- 40%-60% growing
  Joint planting --------------- 50%-70% growing

Growth should be continuous with no open spaces greater than 2 feet in linear systems. Spaces 2 feet or less will fill in without hampering the integrity of the installed living system.

Maintaining the System

After inspection and acceptance of the established system, maintenance requirements should be minor under normal conditions. Maintenance generally consists of light pruning and removal of undesirable vegetation. Heavy pruning may be required to reduce competition for light or stimulate new growth in the project plantings. In many situations, installed soil bioengineering systems become source sites for future harvesting operations. The selective removal of vegetation may be required to eliminate undesirable invading species that should be cut out every 3 to 7 years.

More intensive maintenance will sometimes be required to repair problem areas created by high intensity storms or other unusual conditions. Site washouts should be repaired immediately. Generally, reestablishment should take place for a 1-year period following construction completion and consist of the following practices:

- Replacement of branches in dead unrooted sections
- Soil refilling, branchpacking, and compacting in rills and gullies
- Insect and disease control
- Weed control

Gullies, rills, or damaged sections should be repaired through the use of healthy, live branch cuttings preferably installed during the dormant season. The repair should use the branchpacking system for large breaks and the live gully repair system for breaks up to 2 feet wide and 2 feet deep. If the dormant season has passed, the use of rooted stock may be considered.

Vegetated Structures

Vegetated structures consist of either low walls or revetments (concrete or rock and mortar) at the foot of a slope with plantings on the interposed benches. A structure at the foot of a slope protects the slope against undermining or scouring and provides a slight buttressing effect. In the case of low walls, it allows regrading of the slope face to a more stable angle without excessive retreat at the crest.

Vegetation planted on the crest of the wall and the face of the slope protects against erosion and shallow sloughing. In the case of tiered structures, the roots of woody plants grow into the soil and backfill within the structure, binding them together. The foliage in front covers the structure and enhances its appearance. These systems are not soil bioengineering structures, as their plant materials represent little or no reinforcement value to the structure.

Low wall/slope face plantings

A low retaining structure at the foot of a slope makes it possible to flatten the slope and establish vegetation. Vegetation on the face of the slope protects against both surface erosion and shallow face sliding. Materials and installation-Several basic types of retaining structures can be employed as low walls. The simplest type is a gravity wall that resists lateral earth pressures by its weight or mass. The following types of retaining structures can be classified as gravity walls:

— Masonry and concrete walls

— Crib and bin walls

— Cantilever and counterfort walls

— Reinforced earth and geogrid walls

In addition, each of these can be modified in a variety of ways to fit nearly any condition or requirement.

Tiered Wall/bench Plantings

Description-An alternative to a low wall with face planting is a tiered retaining wall system. This alternative effectively allows vegetation to be planted on slopes that would otherwise be too steep. Shrubs and trees planted on the benches screen the structure behind and lend a more natural appearance while their roots permeate and protect the benches. Virtually any type of retaining structure can be used in a tiered wall system. A tiered wall system provides numerous opportunities for adding vegetative values on steep slopes and embankments.

Cribwalls with Plantings

A cribwall is a structure formed by joining a number of cells together and filling them with soil, gravel, or rock to furnish strength and weight. In crib structures, the members are essentially assembled "log cabin" fashion. The frontal, horizontal members are termed stretchers; the lateral members, headers. The frontal spaces between the stretchers in conventional cribwalls provide openings through which vegetative cuttings can be inserted and established in the crib fill.

References

Fridl, W.F., and P.E. Demetrious., *Biotechnical bank stabilization*, Public Works, 1982.

Gray, D.H., and R. Sotir., "Biotechnical stabilization of a highway cut slope," *J. Geotechnical Eng. Amer. Assoc. Civil Eng.*, Vol 118, No. 9.2, 1992.

Kraebel, C.J., "Erosion control on mountain roads," USDA Circ. No. 380, 43 pp, 1936.

12

Contaminated Soil Treatment Technology

The Engineering, Federal Facilities, and Ground Water Forums, established by EPA professionals in the ten regional offices, are committed to identifying and resolving scientific, technical, and engineering issues impacting the remediation of Superfund sites and corrective action sites under the Resource Conservation and Recovery Act (RCRA). The forums are supported by and advise the Office of Solid Waste and Emergency Response's (OSWER) Technical Support Project, which established Technical Support Centers in laboratories operated by the Office of Research and Development, Office of Radiation Programs, and the Environmental Response Team. The centers work closely with the forums, providing state-of-the-science technical assistance to EPA project managers.

In Situ Technologies

The *in situ* technologies are categorised into three major groups based on the primary mechanism by which treatment is achieved:

— Physical/Chemical Treatment Technologies

— Biological Treatment Technologies

— Thermal Treatment Technologies

Physical/chemical treatment includes soil vapour extraction, solidification/stabilisation, soil flushing, chemical oxidation, and electrokinetic separation. Biological treatment uses microorganisms or vegetation to degrade, remove, or immobilise contamination in soil. Biological technologies include bioventing, phytoremediation, and monitored natural attenuation. Electrical resistivity heating, steam injection and extraction, conductive heating, radio-frequency heating, and vitrification are technologies summarised under thermal treatment.

The principal feature of many in situ treatment technologies is delivery and recovery of fluids or other reactants to the subsurface. The ability to control and monitor the delivery and recovery of these fluids or reactants is central to the effectiveness of *in situ* technologies in treating the contamination. Depending on the subsurface conditions and contaminant characteristics, each *in situ* technology has benefits and limitations on its ability to effectively deliver, control, and recover administered fluids and/or reactants and the contaminants.

For example, soil permeability is an important factor in the delivery of a reactant for chemical oxidation or a gas for bioventing, whereas it is not as important for conductive heating. Consequently, the characterisation of this parameter would generally be more critical for chemical oxidation or bioventing than for conductive heating. The increased use in recent years of several in situ soil treatment technologies, such as chemical oxidation and thermal treatment, has shown that both technologies are a viable option for addressing source zones contaminated by nonaqueous phase liquids (NAPLs).

In addition, greater emphasis is being placed on examining these technologies for their potential synergies as treatment trains to address contamination in the subsurface. This integrated approach has the potential for providing a more effective site remediation.

Physical/Chemical Technologies

Physical/chemical technologies, which represent the most diverse group of remediation technologies, include soil vapour extraction, solidification/stabilisation, oxidation, soil flushing, and electrokinetic separation.

Soil Vapour Extraction (SVE)

In situ soil vapour extraction (SVE) is a remediation technology in which a vacuum is applied to induce a controlled subsurface air flow to remove volatile organic compounds (VOCs) and some semivolatile organic compounds (SVOCs) from the vadose zone to the surface for treatment. The configuration of the system usually involves attaching blowers to extraction wells which are generally constructed with slotted polyvinyl chloride (PVC) to induce airflow through the soil matrix.

The contaminated air is brought to the surface and passed through a vapour/liquid separator to remove any moisture before the air is treated. Treatment is typically done by adsorption (activated carbon), or for more concentrated waste streams, by thermal oxidation systems. The water generated by the liquid separator may also require treatment. When expected concentrations in the air stream are sufficiently high (1,000 to 5,000 parts per million [ppm] or more) for free product recovery for recycling, a stand alone condensation treatment system might be considered. This type of system is generally not used for mixtures of chemicals, and at some point the condenser system will need to be changed out when concentrations drop.

Concrete, asphalt, geomembrane, or other low-permeability covers are often placed over the soil surface to prevent short-circuiting of air flow and to increase the radius of influence of the extraction wells. Replacement air can be introduced into the subsurface by injecting air via a blower or by allowing air to flow into passive injection wells. While vertical wells are the most widely used SVE design method, when the contamination and/or the water table is shallow, horizontal wells or trenches provide better lateral flow and superior formation access.

The SVE process is driven by the partitioning of volatile materials from condensed phases (sorbed on soil particles, dissolved in pore water, or nonaqueous liquid) into the soil gas being drawn through the subsurface. The partitioning is controlled by contaminant and soil properties. These properties include contaminant vapour pressure, Henry's law constant, solubility, soil intrinsic permeability, water content (which should be low, but very dry soils also inhibit contaminant mobilisation), and organic carbon content.

SVE is best suited in well-drained, high-permeability soil (sand and gravel) with a low organic carbon content. Low permeability soil or heterogenous soil with high carbon content are more difficult to treat with SVE and often require amendments, such as pneumatic or hydraulic fracturing. Fracturing allows for high preferential flow paths, but the bulk of the contaminant load still depends upon low flow or diffusion from the competent soil matrix.

Like fracturing, heterogenous subsurfaces provide differential flow paths that result in efficient removal of contaminants in the permeable layers, with the less permeable layers being subject to slow diffusive forces. Rate-limited diffusion in the less permeable soils extends the time needed for remediation; therefore, it may be more efficient to approach these types of sites with a pulsed pumping strategy, in which the blowers are turned off at predetermined effluent concentrations, and the contaminants are allowed to diffuse into the "clean" permeable layers.

After a suitable (site-specific) time, the blowers are turned back on to capture the more concentrated soil vapors. If appropriate, this method can save money on electricity and other costs. When designing an SVE system, DiGiulio and Varadhan advise care in choosing standard radius of influence (ROI) methods to place extraction wells. These methods generally rely on measuring vacuum differentials with distance from the venting well. Vacuum measurements can indicate the direction of a flow gradient, but as the vacuum measured approaches ambient pressures, they may give a false indication and lead to placing wells too far apart.

In addition, vacuum measurements give no information on the effective gas flow through the various subsurface materials. For example, one-dimensional measurements made on layers of sand and silty clay will yield equivalent vacuums, while the effective gas flow is through the sand, with little going through the silty clay. A more relevant approach to well layout is to achieve a pore velocity that exceeds some minimum rate everywhere within the contaminated zone. As the vapour extraction system continues to operate, effluent contaminant concentrations generally become asymptotic (steady-state removal of very low concentrations).

Unless the SVE system is addressing a single contaminant species, measurements of the venting effluent should provide the total mass being removed as well as relative compound concentrations. Speciation data also help in evaluating the system's efficiency. Because the chemicals in a mixture have different chemical/physical properties, they will leave the mixture at different rates; hence, a drop in total concentration does not necessarily mean a drop in available contaminant or system efficiency, but rather exhaustion of certain species. It is also important to test each extraction well in the system individually to determine if the drop is occurring across all wells.

Testing of the header alone may mask wells that have low flow and high concentrations that are being diluted by other wells in the system. Maintaining asymptotic levels over a period of many months is often interpreted as a sign that the SVE effort has been successful and should be shut down; however, as USACE states: "although the decrease of concentrations in the extracted vapour is an indication of the effectiveness of the system, it is certainly not conclusive evidence that the concentrations in the soil have decreased proportionally."

If no rebound is found after shutting the system down for a site-specific determined time, then confirmation sampling should be done. Confirmation sampling can be accomplished with an extensive soil gas survey, continuous soil sampling on a statistically determined grid, or professional judgment with sufficient previous characterisation information gained by use of direct push tools, such as the membrane interface probe or, in the presence of hydrocarbons, by laser-induced fluorescence spectroscopy.

If a site has contaminated groundwater, it should be addressed along with the vadose zone contamination. Often this can be accomplished using a multi-phase extraction (MPE) system to simultaneously remove contaminants from soil and extract contaminated groundwater. A discussion of MPE, which is not within the scope of this document, can be found in U.S. EPA and USACE. The cost of SVE is site-specific and depends in part on the hydrogeology, type and amount of contaminants, and whether the offgas requires treatment.

The FRTR website estimates the cost is between $10 and $40 per cubic yard, with a typical pilot program costing between $10,000 and $40,000. The NAVFAC website provides a $20 to $60 per cubic yard estimate. USACE provides a strategy for estimating costs and a checklist for items to include in the estimate. SVE is a mature, widely used technology, and many vendors are capable of implementing the technology.

Solidification and Stabilisation

Solidification and stabilisation (S/S) refer to closely related technologies that use chemical and/or physical processes to treat radioactive, hazardous, and mixed wastes. Solidification technologies encapsulate the waste to form a solid material. The product of solidification may be a monolithic block, a clay-like material, a granular particulate, or some other physical form commonly considered "solid."

Stabilisation technologies reduce the hazard potential of a waste by converting the contaminants into less soluble, mobile, or toxic forms. The physical nature and handling characteristics of the waste are not necessarily changed by stabilisation. Chemical stabilisation relies on the reduction of contaminant mobility by physical or chemical reactions with the contaminant, rather than the contaminant matrix, as is done with solidification.

The mobility of organic and inorganic compounds can be reduced through various precipitation, complexation, and adsorption reactions. Commonly applied inorganic stabilisation agents include soluble silicates, carbon, phosphates, and sulfur-based binders. Organo-clays have been used to stabilise organic chemicals that are poorly addressed by precipitation and complexation reactions.

The S/S process can be accomplished using either inorganic or polymer binders. The most common inorganic binders are Portland cement, pozzolans (siliceous or aluminous materials that can react with calcium hydroxide to form compounds with cementitious properties), and cement/ pozzolan mixtures. While these binders are effective for a range of inorganic cations and anions, a treatability study should be conducted using on-site soil, contaminants, and groundwater.

In situ chemical stabilisation of inorganics using phosphorus based and other compounds was evaluated in September 1998 under EPA's Superfund Innovative Technology Evaluation Program (SITE). The Soil Rescue and Envirobond™remediation products were applied to a small area of lead-contaminated soil at the Crooksville/Roseville Pottery site in southeastern Ohio. These products chelate the metal ions to reduce mobility. The mean Toxicity Characteristic Leaching Procedure (TCLP) lead concentrations were reduced by more than 99 percent for both products. S/ S treatment of organic contaminants with cementitious formulations is more complex than treatment of inorganic contaminants.

While low levels of organic contaminants can be treated using S/S, many organics will interfere with the hydration process and impede the curing of the solid. Subsurface variations in the concentrations of organics can affect both the leachability and final physical properties of the treated wastes or soil. Thorburg et al. used Portland cement to treat a sediment contaminated with coal tar-derived hydrocarbons. The results showed that the treated sediments leached polycyclic aromatic hydrocarbons (PAHs) and midrange aromatic and aliphatic hydrocarbons at concentrations well above their effective solubliities.

Most cementitious processes are exothermic, and the heat generated by the curing process has the potential to volatilise VOCs. The most significant challenge in applying S/S in situ for contaminated soils is achieving complete and uniform mixing of the binder with the contaminated matrix. Three basic approaches are used for mixing the binder with the matrix:

— Vertical auger mixing

— Shallow in-place mixing

— Injection grouting

Vertical auger mixing requires a system of augers to inject and mix binder into the soil. The treatment depth is limited by the torque required to turn the auger. Current testing indicates a limit of depths to less than 150 feet. The auger diameter, which determines the number of holes that need to be drilled for a given areal extent, can range from several meters for shallow mixing to much smaller diameters for deep mixing.

The need for a smaller diameter auger means more holes will need to be drilled per unit area, which increases the cost for the deeper mixing. If VOCs or mercury are present at the site, the contaminant vapors should be captured and treated. The capture is usually accomplished with a hood that covers the mixing area and conveys the gases to an on-site treatment system. Auger mixing is the most commonly applied method for in situ mixing of S/S reagents with soil.

In-place mixing involves the spreading and mixing of binder reagents with waste by conventional earth-moving equipment, such as draglines, backhoes, or clamshell buckets. A large auger rig can also be employed for in-place mixing. The technology is applicable only to surface or shallow deposits of contamination. A novel form of in-place waste mixing can be used for large areas of heavy-metals contaminated soil.

A lime-stabilised biosolid can be plowed into the contaminated soil, yielding a mixture that reduces toxicity and bioavailability of the heavy metals while providing a soil suitable for supporting vegetation. Injection grouting involves forcing a binder containing dissolved or suspended treatment agents into the formation under pressure, thereby permeating the soil. Grout injection may be applied to contaminated formations lying well below the ground surface. The injected grout cures in place, producing an in situ treated mass.

Polymer binders are thermoplastic or thermosetting. Thermoplastic binders are materials that can be repeatedly melted to a flow state and will harden when cooled. Polyethylene, sulfur polymer, and bitumen are examples of theromoplastic binders. Thermosetting binders are materials that require the combination of several liquid ingredients that, when combined, harden to a solid that cannot be reworked. Thermoplastic binders operate in a temperature range of 120 to 180°C, which could be an issue in soil with high moisture content.

Thermosetting binders operate at ambient temperatures, but they are not amenable to high moisture content. While polymer binders are effective, they may be difficult to use in an in situ setting. S/S has been applied to the remediation of hazardous waste sites for more than 15 years. Experience with the technology, especially the inorganic binders, is abundant. The Army Environmental Policy Institute estimates that in situ S/S of metals using a phosphoric apatite binder costs approximately $46 per ton; using Portland cement for metals costs about $125 per ton; using ammonium modified Portland cement for organics costs about $101 per ton; and using polyethylene costs about $609 per ton.

Chemical Oxidation

Chemical oxidation typically involves reduction/ oxidation (redox) reactions that chemically convert hazardous contaminants to nonhazardous or less toxic compounds that are more stable, less mobile, or inert. Redox reactions involve the transfer of electrons from one chemical to another. Specifically, one reactant is oxidised (loses electrons) and one is reduced (gains electrons). There are several oxidants capable of degrading contaminants.

Commonly used oxidants include potassium or sodium permanganate, Fenton's catalysed hydrogen peroxide, hydrogen peroxide, ozone, and sodium persulfate. Each oxidant has advantages and limitations, and while applicable to soil contamination and some source zone contamination, they have been applied primarily toward remediating groundwater.

The type of delivery system selected depends upon the depth of the contaminants, the physical state of the oxidant (gas, liquid, solid), and its decomposition rate. Backhoes, trenchers, and augers have been used to work liquid and solid oxidants into contaminated soil and sludge. Liquids can be delivered either by gravity through wells and trenches or by injection. For vadose zones, gravity has the drawback of a relatively small area of influence.

Pressurised injection of liquids or gases, either through the screen of a well or the probe of a direct push (DP) rig, will force the oxidant into the formation. The DP rig offers a cost-effective way of delivering the oxidant, and if needed, the hole can be completed as a small diameter well for later injections. Potassium permanganate and other solid phase chemical oxidants have also been added by hydraulic or pneumatic fracturing.

The site stratigraphy plays an important role in the distribution of oxidants. Fine-grained units redirect oxidants to more permeable areas and are difficult to penetrate; hence, they can be the source of rebound later on, as contaminants diffuse out. Long-lived oxidants have the potential to remain active as this diffusion occurs, and they can mitigate some of the potential rebound.

Chemical oxidation usually requires multiple applications. In the special case of nonaqueous phase liquids, oxidants that are in a water-based solution will only be able to react with the dissolved phase of the contaminant, since the two will not mix. This property limits their activity to the oxidant solution/NAPL interface. Cost estimates depend on the heterogeneity of the site subsurface, soil oxidation demand, stability of the oxidant, and type and concentration of the contaminant. Care should be taken when comparing different technologies on a cubic yard basis without considering these site attributes. Cost data can be found in ITRC and Brown. In situ chemical oxidation has been used at a number of sites and is available from a variety of vendors.

Sodium or Potassium Permanganate: Permanganate is a non-specific oxidizer of contaminants with low standard oxidation potential and high SOD. It can be used over a wide range of pH values and does not require a catalyst. Permanganate tends to remain in the subsurface for a long time, allowing for more contaminant contact and the potential of reducing rebound. As permanganate oxidizes organic materials, manganese oxide (MnO_2) forms as a dark brown to black precipitate.

During the treatment of large bodies of NAPL with high concentrations of permanganate, this precipitate may form a coating that reduces contact between oxidant and NAPL. The extent to which this reduction negatively impacts contaminant oxidation has not been quantified. Potassium permanganate has a much lower solubility than sodium and is generally applied at lower concentrations. Commercial-grade permanganates may contain elevated concentrations of heavy metals, and they may lower the pH of the treated zone.

If bioremediation is planned as a polishing step, permanganate will have an adverse effect on microbial activity and may cause a change in microbe distribution. This effect is generally transitory. Also, there is some evidence that permanganates may be inhibitory to Dehalococ-coides ethenogenes, the microbial species that completely dechlorinates tetrachloroethene (PCE) and trichloroethene (TCE).

Fenton's Catalysed Hydrogen Peroxide: Fenton's reagent uses hydrogen peroxide in the presence of ferrous sulfate to generate hydroxyl radicals that are powerful oxidants. The reaction is fast, releases oxygen and heat, and can be difficult to control. Because of the fast reaction, the area of influence around the injection point is small. In conventional application, the reaction needs to take place in an acidified environment, which generally requires the injection of an acid to lower the treatment zone pH to between three and five.

The reaction oxidises the ferrous iron to ferric iron and causes it to precipitate, which can result in a loss of permeability in the soil near the injection point. Over time, the depletion of the ferrous ion can be rate limiting for the process. Chelated iron can be used to preserve the iron in its ferrous state at neutral pH, thus eliminating the acid requirement.

The byproducts of the reaction are relatively benign, and the heat of the reaction may cause favorable desorption or dissolution of contaminants and their subsequent destruction. It also may cause the movement of contaminants away from the treatment zone or allow them to escape to the atmosphere. There are safety concerns with handling Fenton's reagent on the surface, and the potential exists for violent reactions in the subsurface. In many cases there may be sufficient iron or other transition metals in the subsurface to eliminate the need to add ferrous sulfate.

Hydrogen Peroxide: While catalysts can be added to increase oxidation potential, hydrogen peroxide can be used alone to oxidise contaminants. Peroxide oxidation is an exothermic reaction that can generate sufficient heat to boil water. The generation of heat can assist in making contaminants more available for degradation as well as allowing them to escape to the surface. With its high reaction and decomposition rates, hydrogen peroxide is not likely to address contaminants found in low permeability soil. Solid peroxides in slurry form moderate the rate of dissolution and peroxide generation, thereby allowing a more uniform distribution.

Ozone: Ozone, which is one of the stronger oxidants, can be applied as a gas or dissolved in water. As a gas, ozone can directly degrade a number of chemicals in both the dissolved and pure forms, and it provides an oxygen-rich environment for contaminants that degrade under aerobic conditions. It also degrades in water to form radical species, which are highly reactive and non-specific. Ozone may require longer injection times than other oxidants, and vapour control equipment may be needed at the surface. Because of its reactivity, ozone may not be appropriate for slow diffusion into low-permeability soil.

Sodium Persulfate: Persulfate ($S_2O_8^{-2}$) is a strong oxidant with a higher oxidation potential than hydrogen peroxide and a potentially lower SOD than permanganate or peroxide. Persulfate reaction is slow unless placed in the presence of a catalyst, such as ferrous iron, or heated to produce sulfate free radicals that are highly reactive and capable of degrading many organic compounds. At temperatures above 40°C, persulfate becomes especially reactive and can degrade most organics. Like Fenton's reagent, the ferrous iron catalyst will degrade with time and precipitate.

Electrokinetic Separation

Electrokinetic separation is an emerging technology that relies on the application of a low-intensity, direct current through the soil to separate and extract heavy metals, radionuclides, and organic contaminants from unsaturated soil, sludge, and sediment. The current is applied across electrode pairs that have been implanted in the ground on each side of the contaminated soil mass. During electromigration, positively charged chemical species, such as metals, ammonium ions, and some organic compounds, move toward the cathode, and negatively charged chemicals, such as chloride, cyanide, fluoride, nitrate, and negatively-charged organic species, migrate toward the anode.

Electromigration does not require advective flow of pore water for the chemical species to move. In fine-grained soil, the electric current also causes electroosmosis, which is an electrically induced hydraulic flow of ground or soil pore water between the electrodes. This flow can carry neutrally charged species with it. Suspended, charged colloids and miscelles can also move by electrokinetics through the process of electrophoresis. Electrophoresis, in this instance, is similar to electromigration except that the species moving are not single molecules.

Electrolysis reactions create H_2 and OH^- at the cathode and O_2 and H^+ at the anode. These reactions create an acid front near the anode and a base front near the cathode that migrate towards each other. The acid front aids in increasing the mobility of cationic species, but in some soils, it can retard electroosmois. The hydroxide front needs to be controlled to avoid the premature precipitation of some target metal ions. This technology can be applied to contaminant concentration ranges from a few ppm to greater than 10,000 ppm, but may not be effective for treating multiple contaminants that have significantly different concentrations.

The target compounds are either extracted to a recovery system or deposited at the electrode. Surfactants and complexing agents may be used to increase solubility and assist in the movement of the contaminant, although care should be taken when choosing between charged (anionic/ cationic) and neutral surfactants. When electroosmotic flow is from the anode to the cathode, the flow will assist cationic species and retard anionic ones. For the electrokinetics to work, the soil moisture must be conductive and sufficient to allow electromigration but, optimally, not saturated. Removal efficiencies are directly related to the solubility of the target contaminant, its electrical charge, and its concentration relative to other ions or contaminant species.

References

Buol, S.W., F.D. Hole, and R.J. McČracken., *Soil Genesis and Classification,* Ames, Iowa: The Iowa State University Press, 1973, 360 pp.

Brady, N.C., *The Nature and Properties of Soils,* New York: MacMillan, 1974, 639 pp.

Kramer, Steven L., *Geotechnical Earthquake Engineering,* Prentice-Hall, Inc., 1996.

Mitchell, James K. & Soga, K., *Fundamentals of Soil Behavior* 3rd ed., John Wiley & Sons, Inc., 2005.

Rajapakse, Ruwan., *Pile Design and Construction,* 2005.

Bibliography

Brown R. B., "Soil Texture," *Fact Sheet SL-29*, University of Florida, Institute of Food and Agricultural Sciences, Retrieved on 2007-12-02, September 2003.

Buol, S.W., F.D. Hole, and R.J. McCracken., *Soil Genesis and Classification*, Ames, Iowa: The Iowa State University Press, 1973, 360 pp.

Coleman, D.C. and Crossley, D.A. Jr., *Fundamentals of Soil Ecology*, 2nd ed., Academic Press, Burlington, MA, 2003.

Crossley D.A., *Roles of Microflora and fauna in soil systems*, International Symposium on Pesticides in Soils, Feb. 25, 1970, University of Michigan.

Dooley, Alan., Sandboils 101: Corps has experience dealing with common flood danger. US Army Corps of Engineers. Retrieved on 2006-08-29, June, 2006.

Graham, E. R., "An explanation of theory and methods of soil testing," *Missouri Agric. Exp. Stn. Bull.*, 1959, 734.

Huffman, S.A. and K.A. Barbarick., "Soil nitrate analysis by cadmium reduction," *Comm. Soil Sci. Pl. Anal.*, 12(1):79-89, 1981.

Jones, J. A. A., "Soil piping and stream channel initiation", *Water Resources Research* 7 (3): 602 - 610, 1976.

Kiem, R. and Ko¨gel-Knaber, I., "Contribution of lignin and polysaccharides to the refractory carbon pool in C-depleted arable soils," *Soil Biol. Biochem.*, 35, 101–118, 2003.

Luebs, R.E., G. Stanford, and A.D. Scott., "Relation of available potassium to soil moisture," *Soil Sci. Soc. Amer. Proc.*, 1956, 20:45-50.

Mitchell, James K. & Soga, K., *Fundamentals of Soil Behavior* 3rd ed., John Wiley & Sons, Inc., 2005.

Negraa, Christine; Donald S. Rossa and Antonio Lanzirottib, "Oxidizing Behavior of Soil Manganese: Interactions among Abundance, Oxidation State, and pH", *Soil Science Society of America Journal* 1 (69): 97-95. Retrieved on 2006-11-11, 2005.

Poulos, S. J., *Advance Dam Engineering for Design, Construction, and Rehabilitation: Liquefaction Related Phenomena,* Ed. Jansen, R.B, Van Nostrand Reinhold, 1989, pages 292-297.

Rasmussen, P.E. *et al.*, "Crop residue influences on soil carbon and nitrogen in a wheat-fallow system," *Soil Sci. Soc. Am. J.*, 44, 596-600, 1980.

Roth, G.W., D.B. Beegle, R.J. Fox, J.K. Toth, and W.P. Piekielek., "Development of a quicktest kit method to measure soil nitrate," *Comm. Soil Sci. Plant Anal.* 22:191-200, 1991.

Schofield, R. K., and A. W. Taylor., "The measurement of soil pH," *Soil Sci. Soc. Amer. Proc.* 19:164-167, 1955.

Terzaghi, K., Peck, R.B., and Mesri, G., *Soil Mechanics in Engineering Practice,* Third Edition, John Wiley & Sons, Inc. Article 18, 1996, page 135.

Thien, S.J., D.A. Whitney, and D.L. Karlen., "Effect of microwave radiation drying on soil chemical and mineralogical analysis," *Comm. Soil Sci. Plant Anal.*, 1978, 9:231-241.

Voroney, R. P., "The Soil Habitat in Soil Microbiology," *Ecology and Biochemistry*, Eldor A. Paul (ed.), 2006.

Index